Collins

College
ries
d on or befo
halow

D0351816

Student Support
Materials for AQA

AS and A2 Geography

Units 2 and 4

Geographical Skills, Fieldwork
Investigation and Issue
Evaluation

Authors: Geoff

Ser

* 000286939 *

Published by Collins Education
An imprint of HarperCollins Publishers
77-85 Fulham Palace Road
Hammersmith
London
W6 8JB

Browse the complete Collins Education catalogue at
www.collinseducation.com

© HarperCollins Publishers Limited 2011

10 9 8 7 6 5 4 3 2 1

ISBN 978 0 00 741571 7

Geoff Gilchrist and Tania Grigg assert their moral rights to be identified as the authors of this work.

All rights reserved. No part of this publication may be reproduced, stored in a retrieval system, or transmitted in any form or by any means, electronic, mechanical, photocopying, recording or otherwise, without the prior written permission of the Publisher or a licence permitting restricted copying in the United Kingdom issued by the Copyright Licensing Agency Ltd, 90 Tottenham Court Road, London W1T 4LP.

British Library Cataloguing in Publication Data.

A Catalogue record for this publication is available from the British Library.

Commissioning Editor: Lucy McLoughlin

Project Editor: Lucien Jenkins

Design and typesetting: Hedgehog Publishing Ltd

Cover Design: Angela English

Index: Indexing Specialists (UK) Ltd

Production: Simon Moore

Printed and bound by L.E.G.O. S.p.A., Italy

Acknowledgements

Every effort has been made to contact the holders of copyright material, but if any have been inadvertently overlooked the publishers will be pleased to make the necessary arrangements at the first opportunity.

Cover and p. 1 © David Chadwick/istockphoto.com; p. 5 © Ordnance Survey; p. 6 © Collins Geo; p. 7 © Collins Geo; p. 8 © Collins Geo; p. 9 © Crown/Collins Geo; p. 10 © Collins Geo; p. 11 © Collins Geo; p.14 © Crown/www.statistics.gov.uk; p. 15 ©UN/un.org; p. 47 © Friends of Hilbre/www.deeestuary. co.uk; p. 55 © Crown/www.neighbourhood.statistics.gov.uk; p.56 © Crown/www.neighbourhood. statistics.gov.uk; p. 64 Collins Geo; p. 65 Rob Philpott © National Museums Liverpool

All other photos supplied by the authors

Contents

Examiners' notes

For AS and A2 you have to be able to use and read:

- Atlas maps
- Base maps
- Sketch maps
- Ordnance Survey maps at 1:25000 and 1:50000 scales
- Detailed town centre plans

You also have to understand and be able to construct:

- Maps with located proportional symbols
- Maps showing movement, e.g. flow lines, desire lines and trip lines
- Choropleth, isoline and dot maps

Examiners' notes

You are not expected to know the location of every country in the world. However, it is very useful for A-level Geography if you know the location of the continents, major countries and case-study places.

Cartographical skills are those that involve the use of a map. There are two main skills:

- Reading and interpreting maps
- Using maps to present data spatially

Atlas skills

Make sure that you are able to use an atlas. This means you should be able to understand and use lines of **latitude** and lines of **longitude.** They are useful when referring to location. GPS (global positioning systems) devices such as 'sat navs' use latitude and longitude to determine positions.

Familiarize yourself with using the index to locate places. Atlases contain a wealth of maps that can be useful for comparing countries and for collection of secondary data.

Ordnance Survey (OS) maps

To be able to read an OS map correctly you need to practise a variety of skills, described below. If you are unsure of any of these skills, ask your teacher for help. It may also be useful to visit the following website which has lots of advice and help with using OS maps: http://mapzone. ordnancesurvey.co.uk/mapzone/

For A-level Geography you need to be able to understand and use the following:

- Grid references (4– and 6–figure)
- Scale
- Compass direction
- Height and relief (contour patterns)

Fig 1 on page 5 is an Ordnance Survey extract from a 1:25000 scale map, covering the area around Keswick in the Lake District. It is not necessary to know all the symbols off by heart as in an exam you will be supplied with the key. However, it is quicker and easier if you can learn some of them so you won't need to constantly refer to the key. The key has not been included with this extract but you can find it on any 1:25000 OS map.

Sketch maps

These are useful for showing the location of a case study or an area of fieldwork. You do not need to include every detail – decide on your priorities. Every sketch map should include the following:

- Title
- North arrow
- Labels to indicate important features

Maps with proportional symbols

Maps can help you to investigate spatial patterns and compare data between different locations. By using a map with **proportional symbols** it is possible to investigate spatial patterns together with aspects of volume or size of data.

Fig 1 1:25 000 OS map extract of the area around Keswick in the Lake District

What they are: Proportional symbols are symbols that are proportionate in area or volume to the value of data they represent. They can take a variety of forms:

- Squares
- Circles (see **fig 2**) – see p. 20 for help constructing proportional circles
- Pie charts
- Bar graphs

When to use them: They are useful if you want to compare data across different locations. To make drawing the map worthwhile you need at least three different locations. However, if you have too many it will make the exercise very time consuming. Examples of where they are useful include looking at downstream changes along the long profile of a river, or pebble size variations along a stretch of coastline, or population characteristics of cities or countries across the globe.

Examiners' notes

It is important that the location of the proportional symbol is as accurate as possible. To achieve this, place the symbol as near to the location as possible and, if appropriate, use arrows to pinpoint exact locations. Remember to include the scale of the symbols and use a key if necessary.

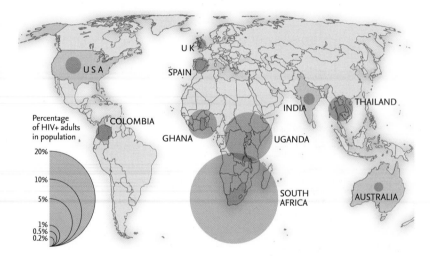

Fig 2
Located proportional circle map to show HIV prevalence in selected countries

Maps showing movement – flow lines, desire lines and trip lines

These are all used on maps to show movement in the form of arrows or lines. They can show direction of movement as well as **density** of movement. They are particularly useful in human geography studies.

Flow lines and **desire lines** both show the volume of movement and, in both cases, the width of the line is proportional to the quantity of movement. The difference between them is that a flow line shows the quantity of movement along an actual route whereas a desire line is drawn from the point of origin to the actual destination and takes no account of the actual route. The worked example on pp. 7-8 shows how to create a flow line map.

Essential notes

Keep the background map as simple as possible so as not to clutter or obscure the flow lines.

Flow line and desire line maps may be useful to show:
- Movement of traffic across a city
- Migration routes

- Tourist destinations
- Origins of visitors or shoppers

Trip lines are drawn to show regular trips or journeys that people make. For example, a map can be drawn to show the places on a map that people commute to from a village location.

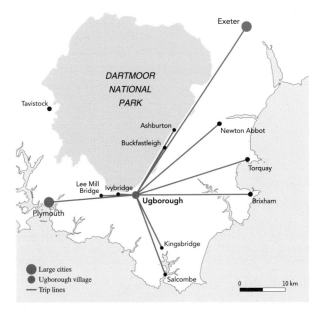

Fig 3
Trip lines of commuter journeys from Ugborough, Devon

How to create a flow line map

A group of students investigating environmental quality across different areas of Plymouth decided to use a flow line map to present their results of a traffic survey.

They collected the data by counting how many vehicles passed them going towards the city centre in a 10-minute period. Their locations were selected strategically so that they stood on a main route in specific wards. They collected the data at the same time at each location.

Road	No. of vehicles			
	Inner-city	**Inner suburb**	**Outer suburb**	**Rural-urban fringe**
A379	298	147	76	23
A386 B3250	233	262	189	87

Table 1
Traffic flow along the A379 and A386 towards Plymouth city centre

As there was a large variation in the figures the students decided to categorize the results into six groups: 1–50, 51–100, 101–150, 151–200, 201–250 and 251–300.

They then worked out a scale so that each category of 50 vehicles represented 1 mm. The flow lines were then drawn according to the scale and in the direction of flow for each location. **Fig 4** shows the final flow line map.

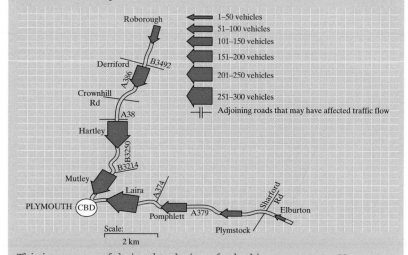

Fig 4
Flow line map showing Plymouth traffic flow

This is a very useful visual technique for looking at patterns. You can clearly see the areas that have a lower traffic flow.

Choropleth maps

A **choropleth map** shows the relative density of a characteristic in an area. It is completed by shading in colour, greyscale or line density to show how the data values change from location to location.

Choropleth maps are visually striking ways of representing data, as patterns are clearly visible. However, there are also limitations of the technique:

- Data is placed in categories or intervals and there may be a large variation within the category.
- It can take time to construct. If you have too few intervals then a large number of locations may have the same shading, making it hard to see a pattern.
- If you have too many intervals it may be difficult to find that many shades and again it may be difficult to see a pattern.
- It assumes the whole area under one class of shading has a uniform density. In other words, it doesn't show variations that may occur within an area.

Choosing intervals or categories: You may be able to identify the intervals easily; however, in practice, it is usually more difficult. Make sure that you don't include the same number twice, e.g. do not use 1–100 and then 100–200. It should be 0–99.9 and then 100–199.99 and so on.

One way of determining the interval is by completing a dispersion diagram (see p. 26). It is then easy to see natural breaks where the boundaries of

Examiners' notes

If you have to complete a choropleth map in an exam make sure you use the exact shading as shown in the key.

the intervals can be placed. Remember not to choose too many or too few categories – ideally, four to six is best.

Choosing the shading: Shading can take many forms. If you are using colour or greyscale then usually you shade from dark to light to represent highest to lowest values. It is better not to use the extremes of black and white. Black often suggests a maximum value and white is often used to represent 'no data'.

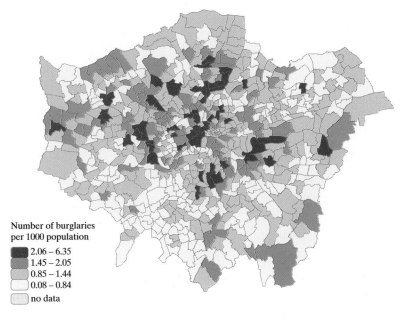

Number of burglaries
per 1000 population

- 2.06 – 6.35
- 1.45 – 2.05
- 0.85 – 1.44
- 0.08 – 0.84
- no data

Fig 5
Choropleth map showing the number of burglaries per 1000 people in London

The choropleth map above shows the density of burglaries in wards across London. A general trend can be seen, showing that the density of burglaries is higher in more central wards. However, there are anomalies; for example, Brunel in the west has a much higher rate than surrounding wards, whereas some central wards such as Barnsbury in Islington have a much lower rate.

Isoline maps

Isolines are lines that are drawn on a map to represent points of equal value. There are many different types and uses of isoline maps:

- Contour lines on Ordnance Survey maps (**fig 1**)
- Pressure lines on a synoptic chart (these are called isobars)
- Temperature (isotherms), for example to show an urban heat island effect
- Travel times for commuters (isochrones)
- River velocities (isovels), for example the cross-section of a meander.

Isolines are very useful for looking at patterns of distribution. However, a large amount of values are needed so they are more suited to group data collection. The more data points you have the more accurate your map will be, although it will be more complicated to draw.

Worked example of isoline map construction

A group of students collected data on the number of pedestrians in Plymouth city centre. They wanted to investigate shopping patterns since the construction of an indoor shopping centre to test the hypothesis: *There will be fewer pedestrians with increasing distance from Drake Circus shopping mall.*

The students collected the data by surveying points around the city centre. They counted how many pedestrians walked past them, in both directions, in a five-minute period. The counts were all conducted between 11:00 and 11:30 a.m. to ensure fairness of the count.

To construct the isoline map, they used a base map of the city centre. They then placed a dot at each survey point and marked on the number of pedestrians recorded at that point. They decided to use intervals of 10 people for each isoline. These were drawn on the map by ensuring they passed through any points of equal value to the isoline. All points of lower value than the isoline lay on one side and points of higher value lay on the other side. The results are shown in **fig 6** below.

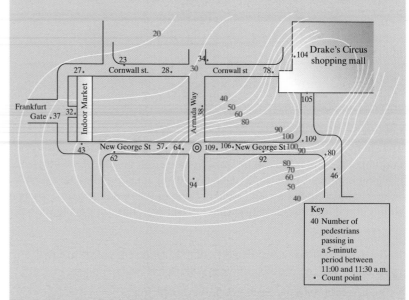

Fig 6
Isoline map showing pedestrian numbers in Plymouth city centre

It can be seen that, in general, there is clearly a decrease in the number of pedestrians with distance from the shopping centre. However, the decrease is not uniform and in some areas there is no evidence that numbers decrease away from Drake Circus. This study would need further investigation.

Dot maps

A **dot map** shows the spatial **distribution** or density of a variable across an area. Each dot represents the same value and therefore, unlike choropleth maps, it is possible to estimate the numbers in a particular area by counting the dots.

Some examples of where dot maps are useful:

- Population density
- Distribution of ethnic groups
- Incidence of disease
- Crime rates

Dot maps are easy to draw but they do have some limitations:

- Where density is very high it is difficult to count the dots, making an estimation of actual values very difficult.
- Scale is often an issue – some areas may have densities well below the dot value so will appear empty – having a value of 0. For this reason, dot maps can be misleading.

The dot map below (**fig 7**) shows the distribution of the Asian population across the city of York. It is easy to see the higher density in central areas. Clear patterns emerge; for example, the highest densities are seen in southern central areas. Lowest densities are seen outside the ring road.

Examiners' notes

When constructing a dot map, the dot should have a high enough value to avoid overcrowding the map with dots but also a low enough value to avoid some areas with low concentrations having no dot.

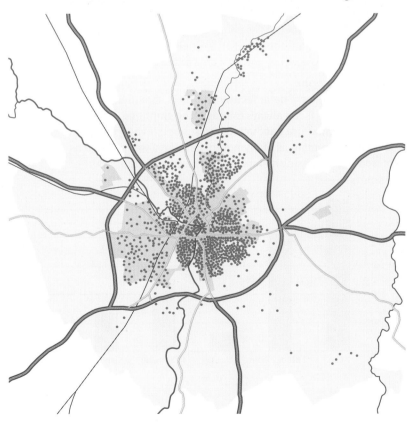

Fig 7
Dot map showing the distribution of the Asian population across the city of York in 2001

Essential notes

This is list of the graphs you must be familiar with:

- Line graphs
- Bar graphs
- Scattergraphs
- Pie charts and proportional divided circles
- Triangular graphs
- Kite and radial diagrams
- Logarithmic scales
- Dispersion diagrams

Examiners' notes

Make sure you take a sharp pencil and ruler into the exam. Many students lose marks because they draw freehand or their pencil is too thick to show accurate plots.

Essential notes

To obtain these results the student counted how many vehicles passed her (in both directions) in a duration of five minutes at 10.00am on weekdays.

For AS and A2 you have to be able to use a variety of graphical skills. This means that you have to be able to construct graphs, choose which type of graph is useful for your data set and be able to justify this choice. You also need to be able to describe patterns and trends seen in graphs and explain the reasons for the patterns.

In this section you will learn how to construct graphs and decide what sort of data each graph is best suited for.

You will use these skills when completing fieldwork as part of your course and throughout Units 1 and 3. You should learn to apply these skills throughout your geography course. You will be examined on these skills particularly in Units 2 and 4A and B but you may also use them in the exams on Units 1 and 3.

Bar graphs

Bar graphs have rectangular columns that rise vertically in proportion to the value they represent. There are several different types:

- Simple
- Comparative
- Compound
- Divergent

Bar graphs are used to show discrete or discontinuous data, e.g. pebble size along a stretch of beach or responses to a questionnaire. The **x-axis** usually represents the categories and the **y-axis** shows the values of the categories either as whole numbers or as percentages of the whole. Bar graphs are simple to understand – just read off the height of the vertical axis to get the value of the category. They can give a basic idea of trend or patterns.

Simple bar graphs

These show one set of data. Technically, there should be a gap between each bar but this is not always the case. The graph (**fig 1**) below shows the result of a traffic count conducted by a student investigating environmental quality in different areas of Plymouth. Each area was selected as representing increasing distance from the city centre.

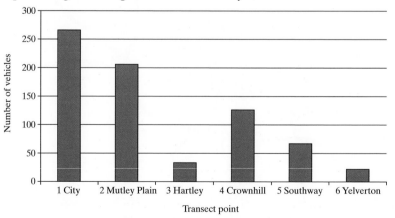

Fig 1
Traffic count survey results for Plymouth study

This graph is useful as it clearly shows a decrease in traffic at greater distances from the city centre. However, it doesn't show actual distance and further study would need to be done to investigate a relationship.

Comparative bar graphs

These show two or more sets of data on the same bar graph and are useful if you want to compare sets of data. For example, you might want to compare monthly rainfall figures for three different years or river discharge during winter and summer at different sites downstream.

In the example below, the student conducting traffic counts in Plymouth has decided to see if the count is different at weekends. She has displayed her results in **fig 2** below.

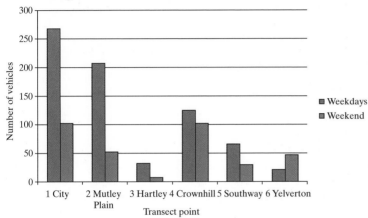

Essential notes

To obtain these results the students conducted the traffic count in the same location but this time at 10.00am on Sundays.

Fig 2
Comparative bar graph showing traffic count results in Plymouth on weekdays and weekends

This graph allows the student to compare the two different times as well as the six locations. This type of graph can be extended to show further comparisons and allows trends to be seen clearly by eye.

Compound bar graphs

These are sometimes also called 'percentage bar graphs'. They allow the individual bars to be broken down into components or percentages. Look at the example on the next page (**fig 3**) to see how they are constructed.

Worked example: constructing a compound bar graph

A student has collected data on the percentage of each plant type across a shingle ridge at Slapton Sands. They collected this data to test the hypothesis: *The number of pioneer species will decrease with distance inland*. They measured the percentage of each plant type using a point quadrat at six different sites moving inland across the shingle ridge.

In constructing the compound bar graph:

- All the values on the y-axis will add up to 100 in this case, as we are looking at the % of each plant type at each transect point.
- Use the same order of categories each time, otherwise the graph will look confusing and be difficult to compare.
- Remember to add on each percentage above the previous one.

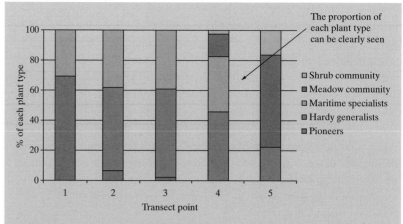

Fig 3
Compound bar graph showing the
percentage of each plant type across
the shingle ridge

This graph makes it very easy to compare the percentages of each
plant type across the ridge and the decline of pioneer species can
be clearly seen. It is simple to read if you remember that they are
constituent parts, e.g. the percentage of maritime specialists at site 4
is 37% and not 83%.

Divergent bar graphs

These show different values either side of the x-axis or either side of the
y-axis. An example of a divergent bar graph either side of the x-axis might
be to show positive and negative values, e.g. a graph showing glacial budget
with accumulation and ablation.

Values can also be spread across the y-axis. This is shown in **fig 4** below.
This shows a population pyramid for Torbay in Devon.

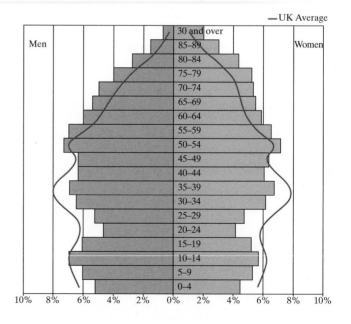

Fig 4
Population pyramid for
Torbay, Devon.

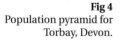

Line graphs

Line graphs are used to show continuous data and are useful for showing trends. For example, they may be used to show change over time such as temperature over a year; or change over distance, for example depth across a river channel or the long profile of a river.

As with bar graphs, line graphs can be simple, compound, comparative or divergent. With simple line graphs, read the actual values off the vertical axis. In a compound line graph the distance between the adjacent points gives the actual value (see **fig 6**). Comparative line graphs show different sets of data such as temperature at different locations over time. The Demographic Transition Model (**fig 5**) is a comparative line graph. Divergent line graphs are similar to divergent bar graphs in that they show data displayed either side of the x- or y-axis.

Examiners' notes

Before you present your data, think carefully what type of graph is most suited for the data. A common error is choosing a line graph to show discontinuous data.

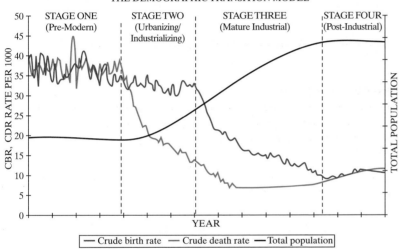

THE DEMOGRAPHIC TRANSITION MODEL

Fig 5
The Demographic Transition Model as seen in the UK

Drawing a line graph:

- Plot the **independent variable** on the x-axis (horizontal) and the **dependent variable** on the y-axis (vertical). For help with this refer to the section on scattergraphs (pp. 17-19).
- Work out the scale before drawing the graph and make sure it fits on your paper.
- Remember to label each axis and to add a title. If you are drawing more than one line, make sure you add a key.

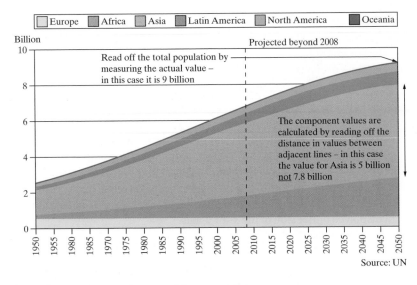

Fig 6
Compound line graph to show world population growth by continent

Drawing cross-sections and long sections

Both of these are types of simple line graphs and are used to show changes in the shape of the land; for example, beach profiles or changes in the depth of the river channel. The horizontal (x) axis shows distance and the vertical (y) scale represents height. You should present your section with a realistic scale and make note of the vertical exaggeration if one is present (see example below).

Examiners' notes

It is important to think carefully about the vertical scale, otherwise you end up with a scale that means very little change is shown if the scale is too small or one that is highly exaggerated where the scale is too big. Too many students present river cross-sections that appear to have depths akin to deep-ocean trenches or beach profiles that look like Olympic ski-jumps.

Table 1
Table of data on river depth

Constructing a river cross-section

The following data has been collected on river depth in order to construct a river cross-section. This has been done to test the hypothesis: *River depth increases with distance downstream.*
This data was collected by firstly measuring the width of the river. The width was then divided by 9 (to give 10 equal points) and then the depth measured at these intervals starting from the left bank, using a metre rule.

Distance/interval (width = 1.89 m)	Left bank	0.21 m	0.42 m	0.63 m	0.84 m	1.05 m	1.26 m	1.47 m	1.68 m	Right bank
Depth (metres)	0.035	0.095	0.146	0.085	0.260	0.350	0.340	0.175	0.170	0.162

- First work out the horizontal scale. In this case the distance is 1.89 m. However, if you plan to draw further cross-sections, start with the widest river as you need to use the same scale for all cross-sections if you want to compare them.
- Work out the vertical scale – in this case a vertical exaggeration of 2 is acceptable. So if your horizontal scale is 1 cm to 10 cm and your vertical scale is 1 cm to 5 cm, then the vertical exaggeration is 2.
- As this is channel depth, remember you are working downwards from the horizontal line! In other words, the surface of the channel

is 0 and you draw your vertical scale increasing with depth (see **fig 7** below).

- Next draw a line horizontally from 0m on the y-axis to represent the surface of the channel.
- Starting at 0 on the x-axis, plot each depth at its corresponding location.
- Join up your points with a freehand line – you can now see the shape of the river bed.

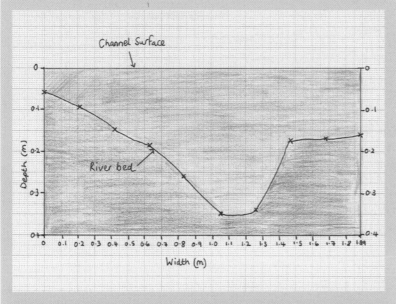

Fig 7
River cross-section

Examiners' notes

Although you can construct a variety of graphs using ICT software such as Excel, it is important that you are able to draw graphs by hand so that you fully understand how to construct them. Cross-sections are an example of graphs that are easier to draw by hand as Excel will not construct them unless you convert your depth value to negatives and even then you may have problems getting the scale right.

To develop this further and provide evidence for the hypothesis you would need to construct river cross-sections at different points with distance downstream.

Scattergraphs

Scattergraphs are used to show the relationship between two variables; for example, in a river study, hydraulic radius and velocity. The relationship is made clearer by adding a line of best fit. This can show whether the relationship has a positive, negative or no **correlation** (**fig 8**).

- Scattergraphs are useful when you have two sets of data.
- You cannot show more than two sets of data on a scattergraph.
- You can identify trends and patterns very clearly by eye.
- **Anomalies** (or residuals) are easily visible.
- The correlation in scattergraphs can be further analyzed by using a statistical test such as Spearman's rank (see p. 28).

Work through the following example to understand how to draw a scattergraph.

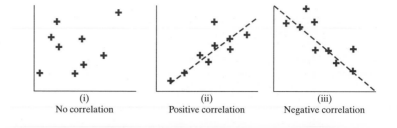

Fig 8
Scattergraphs showing positive, negative or no correlation

(i) No correlation   (ii) Positive correlation   (iii) Negative correlation

Using scattergraphs to present data

The following data on velocity and hydraulic radius has been collected from 10 sites along a river. In order to see if there is a relationship between the two variables a scattergraph has been drawn.

Table 2
Table of data collected from the river study

Site no.	1	2	3	4	5	6	7	8	9	10
Velocity (m/s)	0.093	0.005	0.168	0.187	0.212	0.097	0.260	0.390	0.502	0.381
Hydraulic radius	0.065	0.178	0.162	0.128	0.216	0.164	0.225	0.284	0.314	0.218

These are the steps taken to construct the scattergraph:

- First decide which data is the independent variable and which is the dependent variable. In this example the independent variable is hydraulic radius and the dependent variable is velocity. This is because velocity should be affected by hydraulic radius as the more efficient (larger hydraulic radius) the river is, the faster the water should flow as there will be less friction with the banks and bed of the river.
- Plot the independent variable on the horizontal x-axis and the dependent variable on the vertical y-axis.
- Find a workable scale for each axis. The lines of axis should be similar in length.
- Plot your points at the correct coordinates.
- Look for the general trend or correlation. The correlation is unlikely to be perfect – try to see if there is a pattern. If the crosses, or points, are completely random this means there is no correlation. However, if you can see a trend you are ready to try and add a line of best fit.
- The line of best fit is usually a straight line drawn to represent the general trend of the data. Do this by eye, drawing a straight line through the data. There should be an equal number of points either side of the line. Discount any obvious anomalies before drawing the line (you could highlight these by placing a ring around them). Look for points that are clustered along a general trend. Note: the line of best fit does not have to start at the point of origin (i.e. 0,0).

Essential notes

Velocity was measured in metres per second using a flow meter. Hydraulic radius is a measure of efficiency and is calculated by dividing the cross-sectional area by the wetted perimeter.

In the example below there are seven points either side of the line with two points on the line. The trend in this case is clear.

You should now be able to describe the correlation. Remember to refer to any anomalies.

In this case there is a positive correlation – as hydraulic radius increases, velocity increases. There is only one anomaly where velocity is much lower than the trend line suggests.

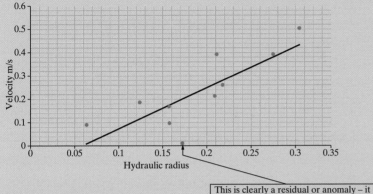

This is clearly a residual or anomaly – it lies some way from the line of best fit – velocity is far slower than the hydraulic radius suggests it should be

Fig 9
Scattergraph to show the relationship between hydraulic radius and velocity

The graph above shows a positive correlation: as hydraulic radius increases, velocity increases. To test this further you could conduct a statistical test such as Spearman's rank result line (see p. 28). This would analyze the strength of the correlation and test the significance of your results.

Pie charts and proportional divided circles

Pie charts are used to compare categories within a data set. A pie chart displays segments of data according to the share of the total value of the data, e.g. categories of pebble shape in a river study. **Proportional divided circles** show segments but they also represent size, as the area of the circle is proportional to the value of the whole data set. They are useful when comparing different data sets.

Pie charts are visually effective as it is easy to see the relative proportions for each segment. However, it is difficult to work out actual percentages and some segments may be too small to see. Proportional divided circles are also visually clear but they can overemphasize large values and therefore smaller values are not as clear.

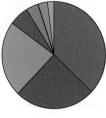

- Oil
- Coal
- Gas
- Nuclear
- Biomass
- Hydroelectric power
- Other renewables

Fig 10
Global energy usage, 2005

Examiners' notes

It is very important to take a protractor into your exam. If you have to draw a pie chart you need to make sure it is accurate.

Essential notes

Formula for constructing proportional divided circles:

$$r = \sqrt{\frac{V}{\pi}}$$

where:
V = value you want the total pie chart to represent
π = 3.142
r = radius of the pie chart

Essential notes

Before you draw your circles you need to decide the unit size of the radius. In this example the student has used 1 mm. However, if the values were too small or too large you would have to adjust this. Make sure you include the scale with your circles.

Constructing a pie chart

- First convert the data into percentages.
- Next convert the percentage for each category into degrees of the circle. As there are 360° in a circle each 1% is worth 3.6°, so therefore multiply each percentage by 3.6.
- Draw your circle and then draw a line from the centre of the circle to the top – this is your starting point. Using a protractor, measure out the degrees for the first category and draw a line from the centre to the point on the circle you have measured. From this point measure the second category in degrees and so on.
- Usually you start with the largest category and move round in a clockwise direction to the smallest but there are no rules concerning this.
- If you are drawing more than one pie chart, make sure you draw the segments in the same order.
- Remember to add a title and a key to show what each segment represents. If appropriate you can add the percentage value to each segment.

Constructing proportional divided circles

Proportional divided circles are useful if you want to display and compare the size of data as well as categorize it. You need to draw more than one for them to be of any value. An example of how to draw them is below. The basic principle is the same as drawing pie charts but this time you draw the radius of the circle so that it is proportional to the value of the data. If you place the proportional circles on a map, you can introduce a spatial element to your presentation.

Worked example of proportional divided circle

The following data (**Table 3**) was collected from a river study. The student is testing the hypothesis: *Pebbles become more rounded and smaller with distance downstream*. In order to present the data, the student has decided to use proportional divided circles as he can show both shape and size on the same graph. It will be clear visually whether there is a larger proportion of rounded pebbles further downstream and whether the size of the proportional circle gets smaller. Thirty pebbles were collected from three sites downstream. At each site the longest axis of the pebbles was measured and the shape of each pebble was assessed using Power's index of roundness. To present the data, the student worked out the average pebble length and converted this using the formula for proportional circles to calculate the radius of each circle. The percentages of each category of roundness were then calculated and converted to degrees by multiplying the percentages by 3.6.

Roundness	Very angular	Angular	Sub-angular	Sub-rounded	Rounded	Well-rounded	Average length
Site 1– upper course	2	14	23	1	0	0	1161 mm Site 1: r = 19.2 mm
Site 2 – middle course	0	5	4	11	8	6	957 mm Site 1: r = 17.5 mm
Site 3 – lower course	1	3	5	14	4	3	496 mm Site 1: r = 12.6 mm

Table 3
Table of results for pebble size and shape

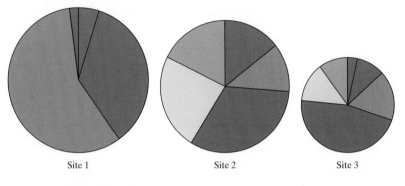

Site 1   Site 2   Site 3

■ Very angular   ■ Angular   ■ Sub-angular   ■ Sub-rounded   ☐ Rounded   ■ Well-rounded

Fig 11
Proportional divided circles of pebble size and shape

It can be seen from the proportional circles that pebble size does decrease from site 1 to 3. Pebble shape is less clear and would need further investigation. More sites would improve this investigation.

Triangular graphs

Triangular graphs, as the name suggests, are graphs that are constructed in the form of an equilateral triangle (**fig 12**). They can only be used if the conditions below apply, and for these reasons they are fairly limited in their use:

- The data must be able to be divided into three component parts.
- The data must be in the form of percentages.
- The percentages must total 100.

Each side of the triangle represents one axis and one component and measures from 0 to 100%. From each axis, lines are drawn at 60° angles to carry the values.

Triangular graphs are useful because you can show a large amount of data on one graph. As with pie charts it is easy to see relative proportions and identify the dominant variable. However, triangular graphs can be difficult to interpret and care must be taken not to make errors reading off the incorrect axis. On the other hand, they have an in-built checking system as all values must total 100. If the total is not 100, an error has been made.

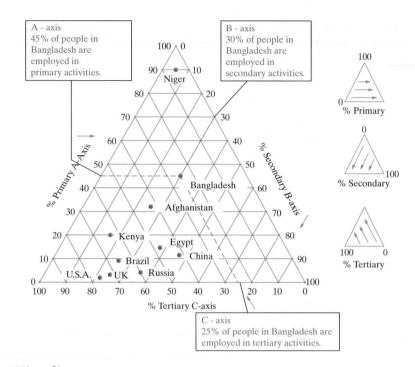

Fig 12
Triangular graph showing
employment sectors

Kite diagrams

Kite diagrams are a useful way of showing changes in vegetation over distance, for example across a sand dune in a study of succession. One axis represents distance across a transect and the other represents individual plant species (it doesn't matter which way round you do them). This axis shows the concentration of that plant species at particular points along the transect.

To construct a kite diagram, follow the steps below using the data in **Table 4**. This data has been gathered from a succession study on a shingle ridge (see p. 14, compound bar graphs). The plants in each quadrat have been grouped into categories. A kite diagram has been drawn to show the change in plant types across the shingle ridge.

Transect distance (m)	1	2	3	4	5	6	7	8	9	10
Hogweed				2.5		5	5	2.5		
Dandelion			2.5	2.5		12.5				
Sea carrot	2.5	12.5								
Sorrel				7.5	2.5	5		7.5		
Ivy							7.5		6	1
Gorse						20	17.5	20		

Table 4
Table showing results of quadrat survey of plant type. Numbers refer to percentage vegetation cover

Step 1: The x-axis represents distance (metres) across the axis.

Step 2: The y-axis represents the six plant types. Make sure there is room to cover all the plant species – if there are too many, consider grouping them.

Step 3: Plot your data and then join up the dots. Note: the amount is plotted either side of 0, to give the symmetrical shape. Therefore divide the value by 2 to give the amount plotted each side of the line. If no species exists at that point draw a thin line or leave it blank. Remember to label the axes and give your diagram a title (see caption for **fig 13** below).

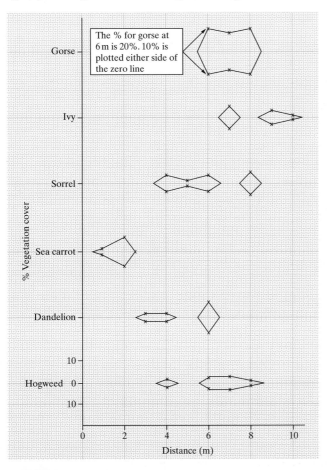

The % for gorse at 6 m is 20%. 10% is plotted either side of the zero line

Fig 13
Kite diagram showing succession on a shingle ridge

Radial diagrams

Radial diagrams are useful for data that has a directional component; for example, wind direction or orientation of pebbles in glacial till. They can also be used in human geography studies to show pedestrian or traffic flow over a period of time.

The radius is used to record frequency or percentage. The scale around the circumference will depend on the data, for example:

- A wind rose will have compass points around the circumference.
- Glacial till or orientation of corries (see **fig 14**) will have bearings in degrees. It is usual to group the bearings into intervals (e.g. 0–30°).
- Pedestrian or traffic flows may have a time element (e.g. a 24–hour period).

Examiners' notes

Radial diagrams can only be used for data where the scale around the circumference is continuous, such as a time sequence or compass bearings.

Worked example for constructing a radial graph

The table below shows the orientation of corries (cirques) on the Isle of Arran. These have been plotted onto a radial diagram (**fig 14**). The circumference shows bearings grouped into 45° intervals and the radius represents frequency.

Orientation of corries on the Isle of Arran (degrees)														
05	05	10	55	15	30	95	05	185	70	120	40	30	115	110

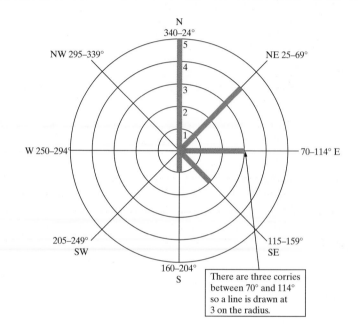

Fig 14
Radial diagram showing the orientation of corries on the Isle of Arran

There are three corries between 70° and 114° so a line is drawn at 3 on the radius.

Logarithmic graphs

Logarithmic graphs are a form of line graph. In an arithmetic line graph the scale increases by equal amounts. **Logarithmic scales** differ from this in that the scales are divided into a number of cycles, each representing a 10-fold increase. An example of this principal is: the first cycle ranges 0.1–10, the second cycle will be 10–100, the third 100–1 000, and so on. **Fig 15** shows an example of logarithmic graph paper.

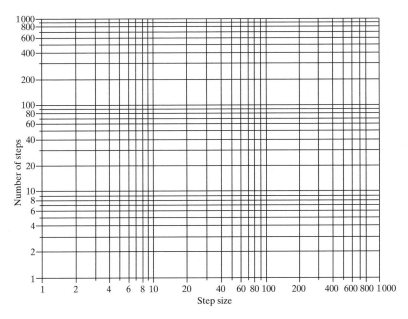

Fig 15
Logarithmic graph paper with a
logarithmic scale on both axes (log-log)

Examples of the use of logarithmic scales include the Richter scale used to measure earthquakes and the Hjulstrom curve (**fig 16**) used in river studies.

Both the x- and y-axis scales can be logarithmic and are called log-log graphs. If only one axis is logarithmic, then it is known as a semi-log graph. Semi-log graphs are useful if looking at change over time. Time is placed on the x-axis and is linear and the corresponding y-axis is logarithmic. This is particularly useful if the data has a large range of values. However, you cannot use positive and negative values on the same graph.

Logarithmic graphs are also useful for showing rates of change, e.g. population change – the steeper the line the faster the rate of change.

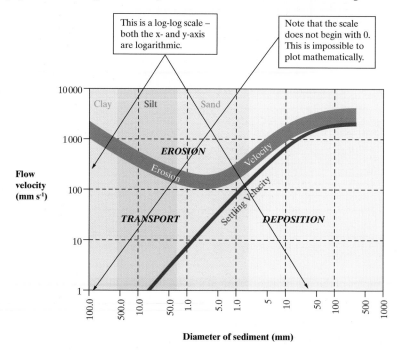

This is a log-log scale – both the x- and y-axis are logarithmic.

Note that the scale does not begin with 0. This is impossible to plot mathematically.

Fig 16
Hjulstrom curve – an example of a logarithmic scale on both axes

Flow velocity (mm s⁻¹)

Diameter of sediment (mm)

Essential notes

Dispersion diagrams can be combined with measures of central tendency. For example, the upper and lower quartiles can be plotted on a dispersion diagram and the interquartile range calculated.

Dispersion diagrams

Dispersion diagrams are used to show the distribution of data. The x-axis is usually very narrow, often representing location or a point in time. The y-axis is longer and represents all the values in the data set. The range of the data and any clustering are therefore clearly visible. Dispersion diagrams are useful for determining the intervals for choropleth maps.

Use a dispersion diagram when:

- There are at least 10 values in the data set.
- You want to see how the data is spread, i.e. the range and degree of clustering.
- You can compare more than one data set as long as the scale is the same. Note the scale does not have to begin at 0.

The dispersion diagram on the next page shows the data used in the pebble study on pages 20–21.

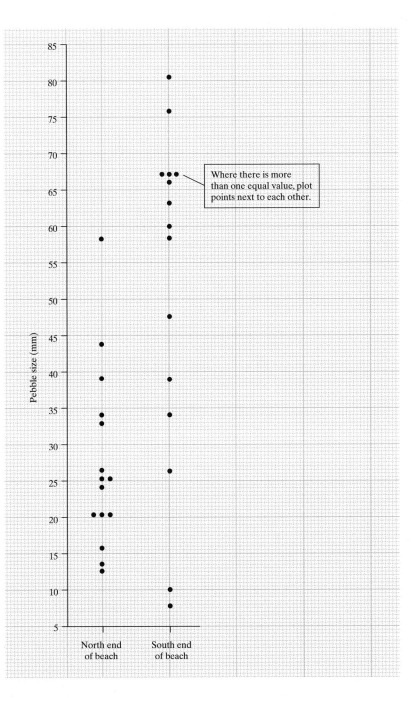

Where there is more than one equal value, plot points next to each other.

Fig 17
Dispersion diagram of pebble size and shape

Essential notes

Statistical skills at AS:

- Measures of central tendency – mean, mode and median
- Measures of dispersion – interquartile range and standard deviation
- Spearman's rank correlation

At A2:

- As above but also comparative tests – Mann-Whitney U and chi-squared

In this chapter you will learn how to use statistical skills and when to apply them. Each skill has a worked example for you to study, to help you apply the skill to your own data.

Spearman's rank result line

Before reading this section, refer back to the section on scattergraphs (pp. 17–19). Spearman's rank is another way of testing a relationship. For example, if you draw a scattergraph you can see by eye if there is a relationship, but you will probably not be able to clearly assess the strength of the relationship as many points may be some distance from the line of best fit.

Spearman's rank correlation is used to test the strength of the relationship between two sets of data, providing you with a numerical value. This is an example of objective data. Once you have this figure you can then test its significance – this means the likelihood of your results occurring by chance.

When can you use Spearman's rank?

The test can be used with any set of raw data or percentages, but it is only suitable if all the following criteria apply:

- There are two data sets which you believe may or may not be related, e.g. hydraulic radius and velocity.
- At least 10 pairs of data are available to be used.
- The test is limited to 30 pairs – more than this makes the exercise unwieldy.

A worked example is shown below. Once you have completed the table and have calculated your answer you should have a figure between –1 and +1. This indicates the strength and type of relationship:

- A figure clearly above 0 indicates a positive relationship (i.e. as one set of data increases so does the other). Refer back to the scattergraphs on p. 18.
- A figure clearly below 0 indicates a negative relationship (i.e. as one set of data increases the other decreases).
- A figure close to 0 means there is no relationship and you would accept the **null hypothesis**.

Examiners' notes

You will not be expected to learn the formula for Spearman's rank or any other statistical test. The formula will be given and you may be expected to complete part of the test and be able to interpret the result. You should also know what the strengths and weaknesses of the test are.

Spearman's rank correlation test	
Strengths	**Weaknesses**
<ul><li>It gives you objective data.</li><li>It enables you to demonstrate a clear relationship between two sets of data.</li><li>You can state whether the relationship is significant or if your results were just a fluke.</li><li>It is less sensitive to anomalies in data as each piece of data is ranked – large differences could only be one rank different.</li></ul>	<ul><li>It does not tell you whether there is a causal link (i.e. that one change leads to a change in another), just that a relationship exists.</li><li>Too many 'tied ranks' can affect the validity of the test.</li><li>It could be subject to human error, e.g. inaccurate calculations.</li></ul>

Case study: Investigating plant succession on a shingle ridge
Students have collected data on the changing abiotic factors across the shingle ridge at Slapton Sands. They want to complete a statistical test to help prove their hypothesis:

Plant height increases as soil depth increases.

As there are two variables (plant height and soil depth) and they have 12 pairs of data they have decided to use Spearman's rank correlation test.

Spearman's rank should always begin with the assumption that there is no relationship – this is called the **null hypothesis**. Always begin your test by writing out the null hypothesis as well as your chosen hypothesis.

Null hypothesis = There is no relationship between plant height and soil depth.

Raw data:

Site	1	2	3	4	5	6	7	8	9	10	11	12
Soil depth (cm)	0.0	3.2	3.6	1.9	10.1	15.2	20.2	23.8	32.0	32.0	34.1	37.4
Plant height (cm)	4.0	1.5	6.0	11.5	22.0	65.0	92.0	103.0	129.0	187.4	156.6	189.3

Table 1
Raw data for plant height and soil depth

A scattergraph has already been drawn for this data.

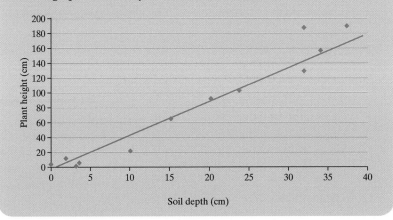

Fig 1
Scattergraph of soil depth and plant height

It is clear that there is a positive relationship but Spearman's rank will test this further and see whether the relationship is significant or could have occurred by chance.

The students carry out the following steps.

Step 1: Complete columns A and B by adding the raw data. (Note: make sure you understand which is the independent variable (IV) and which is the dependent variable (DV).) In this case the students have decided that the DV is plant height because the plant should grow higher if more soil is available to provide nutrients and provide an anchor for roots.

Examiners' notes

It doesn't matter whether you rank your results from highest to lowest or lowest to highest but which ever way you do it **you must do the same for both sets of data**.

Step 2: Rank these two sets of data – the students rank the IV first from deepest soil to shallowest soil (shown in column C). Where there is a tie (in this case sites 9 and 10 for soil depth) they need to work out the average rank (in this case it would have been ranks 3 and 4; add these together and divide by 2: the result is 3.5). The DV is then ranked in the same way (i.e. highest to lowest) as shown in column D.

Step 3: Calculate the difference between the ranks in column E by subtracting the second rank from the first (subtract the value in D from the value in C).

Step 4: Square the difference (this eliminates negative values) and add to column F.

Step 5: You are now ready to start calculating. First work out the total of column F to give a figure for Σd^2 (= sum of). Then multiply this figure by 6.

	A	B	C	D	E	F
n	IV	DV	Rank IV	Rank DV	Rank difference	Difference squared
1	0.0	4.0	12	11	1	1
2	3.2	1.5	10	12	−2	4
3	3.6	6.0	9	10	1	1
4	1.9	11.5	11	9	2	4
5	10.1	22.0	8	8	0	0
6	15.2	65.0	7	7	0	0
7	20.2	92.0	6	6	0	0
8	23.8	103.0	5	5	0	0
9	32.0	129.0	3.5	4	−0.5	0.25
10	32.0	187.4	3.5	2	1.5	2.25
11	34.1	156.6	2	3	−1	1
12	37.4	189.3	1	1	0	0
					$\Sigma D^2 =$	14
					$6\Sigma D^2 =$	84

Table 2
Data ranking of soil depth

IV = independent variable

DV = dependent variable

D = difference in the ranks

n = number of pairs

Σ = sum of

R_s = Spearman's rank

The students are now able to use the Spearman's rank correlation test formula: R_s

$$R_s = 1 - \frac{6 \times \sum d^2}{n^3 - n}$$

In this case they should have the following calculations:

$$R_s = 1 - \frac{84}{12^3 - 12} \qquad R_s = 1 - \frac{84}{1716}$$

Remember that the result of the equation must be subtracted from 1.

So the final calculation is: $R_s = 1 - 0.049$. Therefore $R_s = 0.951$

If the answer is not between −1 and + 1 you have gone wrong and need to start again.

So what does the result of 0.951 mean?

Place the result on a line like the one below:

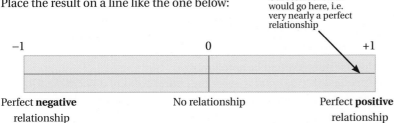

Our result of 0.951 would go here, i.e. very nearly a perfect relationship

−1   0   +1

Perfect **negative** relationship   No relationship   Perfect **positive** relationship

Fig 2
Spearman's rank parameters

Step 6: The students now need to do another test to check whether their result could have occurred by chance – i.e. how significant is the result? This means they have to compare their result with a table of **critical values** (**table 3**). They first look at the number of pairs of data they have – in this case there are 12. They then need to decide which significance level they are going to use. For geographical purposes, you would usually use the **0.05 significance level**. This means that there is a 5 in 100 chance of the results occurring by chance. Or, to put it another way, if other researchers completed the same experiment, 95 out of 100 would get the same result – therefore there is a significant relationship.

The students then need to see whether their R_s result is above the critical value for the number of pairs you have. If the R_s value is below the critical value they must accept the null hypothesis – i.e. they cannot be sure that their relationship is significant.

We can see that in our example the R_s value of 0.951 is well above the 0.05 significance level of + 0.506. In this case it is also well above the 0.1 significance level of + 0.712. This means that there is a very low (1 in 100) chance of the results occuring by chance and we can reject the null hypothesis.

Table 3
Critical values for R_s

n	0.05 (95%) Significance level	0.01 (99%) Significance level
10	+/− 0.564	+/− 0.746
12	0.506	0.712
14	0.456	0.645
16	0.425	0.601
18	0.399	0.564
20	0.377	0.534
22	0.359	0.508
24	0.343	0.485
26	0.329	0.465
28	0.317	0.448
30	0.306	0.432

Examiners' notes

Now have a go at completing a Spearman's rank test for your own data. Follow through the worked example substituting it with your own data. Ask your teacher for help if you are unsure at any stage.

Step 7: You have now completed the Spearman's rank correlation test. Write a statement to summarize your findings.

For these results it would be as follows:
The R_s value was 0.951 and is above the critical value of 0.712 at the 0.01 significance level. I can therefore reject my null hypothesis and accept that there is a strong relationship between plant height and soil depth on the shingle ridge and it is highly significant.

Measuring central tendency: mean, mode and median

Measuring **central tendency** is a measure of the 'middle' value of the data set. There are three ways of measuring central tendency:
- Mean
- Mode
- Median

Examiners' notes

These are some of the easiest statistical skills but you must understand the difference between mean, mode and median. If you muddle up the terms you will lose marks.

These techniques are very useful to geographers, enabling us to summarize a data set by giving the mid-value or most frequently occurring data. They can also be used as part of more complex techniques such as **interquartile range**.

Mean

The **mean** (sometimes called the average) is calculated by adding up all the values in a data set and dividing the total sum by the number of values in the data set.

The mean is particularly useful if the data has a small **range**. However, if the range is large then the mean will be heavily influenced by the extreme values and could give a distorted picture.

Mode

The mode is the value that occurs most frequently in a set of data. You need to know all values before calculating the mode. Mode is of no use if there are no repeating values. There may be more than one mode – this is called 'bi-modal'.

Essential notes

Mean formula:

$$\bar{x} = \frac{\Sigma x}{n}$$

What does this mean?

$\bar{x}$ = arithmetic mean

Σ = sum of

x = observed values

n = number in the sample

Mode is often useful when classifying data, e.g. Power's index of roundness of pebbles in a river study. It is useful to see which classification occurs most frequently. This is called modal class.

Median

This is the middle value in a data set. The data needs to be rank-ordered before you can calculate the **median**.

The median value often needs to be supported by other techniques such as interquartile range (IQR). However, unlike the mean, the median is not affected by extreme values.

Now work through the example below so that you can practise calculating measures of central tendency.

Worked example of mean, mode and median

The aim of the study is to study coastal processes. As part of this aim, a student has collected data on pebble sizes from two sites along the beach at Slapton Sands in South Devon. Fifteen pebbles were measured, at each site, along the A-axis. The results are recorded in the table below in rank order:

Rank	1	2	3	4	5	6	7	8	9	10	11	12	13	14	15
North end of beach	58	43	38	33	32	25	24	24	23	19	19	19	14	12	11
South end of beach	81	76	67	67	67	66	63	60	58	47	38	33	25	8	6

Calculate the mean, mode and median values for each end of the beach. Compare your answers with the ones below.

	North end	South end
Mean	26.3	50.8
Mode	19	67
Median	24	60

Check that your calculations are the same. If there is difference, read through the section again and check you haven't made any errors.

None of these measures gives an accurate picture of the distribution of data. On their own they are of limited value. However, it can be seen from the above example that some judgements can be made. In all three measures the central tendency for pebble size is larger at the southern end of the beach. The average pebble size is much larger at the southern end. The spread of data is quite large, particularly at the southern end, and the extreme low values of 8 and 6 are making the mean value lower than the other two values.

To improve the usefulness of the above calculations, measures of the dispersion or variability of the data should also be calculated.

Essential notes

Median formula

If there is an odd number of values, perform the following calculation to work out the median value:

$$\frac{n+1}{2}$$ (n = number of values in the data set)

Therefore, if you have 23 values in the data set the median will be the 12th value in the rank order.

If the number of values is even, the median is the mean of the middle two values. So if there are 24 values add the values for the 12th and 13th positions and divide by 2.

Table 4
Pebble size (A-axis measured in mm)

Table 5
Central tendency calculations

Measuring dispersion – range and interquartile range

These techniques are used to measure the spread of data. Range and interquartile range allow you to analyze your data in more depth, looking at how spread or dispersed the data is around the mean or median.

Range

This is simply the difference between the highest value and the lowest value. It gives you a basic idea of the spread of data but like the mean it is affected by extreme values. An anomaly therefore can give a false picture.

Worked example of range

Referring back to **table 4**, the range is worked out as follows:

Northern end of beach:
Highest value = 58   **Range: 58 – 11 = 47**
Lowest value = 11

Southern end of beach
Highest value = 81   **Range: 81 – 6 = 75**
Lowest value = 6

Therefore, we can see that the southern end has a much larger range. However, this result is affected by the anomalies of 81 and 6.

Interquartile range

The interquartile range is worked out by ranking the data (highest to lowest) and placing the data into quarters, or quartiles. The top 25% of the data is placed in the upper quartile (UQ) and the bottom 25% is placed in the lower quartile (LQ). The interquartile range or IQR is the difference between the 25% and 75% values.

1st quartile	The boundary between the 1st and 2nd quartiles is called the upper quartile.
2nd quartile	This is the **interquartile range**.
3rd quartile	
4th quartile	The boundary between the 3rd and 4th quartiles is called the lower quartile.

The interquartile range is more useful than the range in indicating the spread of data, as it takes away any extreme values (i.e. those occurring in the UQ 1st quartile and LQ 4th quartile) and considers the spread of the middle 50% of the data around the median or middle value.

The IQR has a formula to work it out. Look at the worked example on the next page to see how to calculate the IQR.

Worked example of interquartile range

- Refer back to our data on pebble sizes. The data is already in rank order, ranked from **highest to lowest.**
- Next find the upper quartile (UQ) by using the formula. In this case the number in the data set is 15, so n = 15. Therefore, using the formula:

$$\frac{15 + 1}{4} = 4$$

NOTE: The answer is not 4. It is the value in the 4th position.

So the upper quartile for the north end = 33 and for the south end = 67.

- Now calculate the lower quartile (LQ):

$$\frac{15 + 1}{4} \times 3 = 12$$

Remember the answer is not 12. It is the 12th position.

The LQ for the north end = 19 and for the south end = 33.

- You are now able to determine the interquartile range using the formula:

UQ – LQ

The IQR for the north end is 33 – 19 = 14

The IQR for the south end is 67 – 33 = 34

This shows us that there is now only a small variation around the median values. There is less variation at the northern end, perhaps suggesting that there is less variation in pebble size, whereas the pebbles are less well sorted at the southern end. However, this would need further investigation. In terms of a comparison between the two sites it would suggest that the pebbles are smaller, and more uniform in size, at the northern end.

Measuring dispersion – standard deviation

Standard deviation is a measure of the degree of **dispersion**. The interquartile range will tell you how clustered the data is around the median value; standard deviation is simply another method of examining the spread of data but this time around the mean.

Two sets of data could have the same mean but have a very different spread of data. Standard deviation will tell you the extent of this – in other words, how reliable the mean is. A low standard deviation indicates that the data points tend to be very close to the mean, whereas high standard deviation indicates that the data is spread out over a large range of values and the mean is less reliable as there is obviously a lot of variation in the sample.

Essential notes

Interquartile range formulae

Upper quartile (UQ) = $\frac{n + 1}{4}$

Lower quartile (LQ) = $\frac{n + 1}{4} \times 3$

Interquartile range (IQR) = UQ – LQ

where: n = number of items in the data set

Examiners' notes

An important point to consider here is sample size. In any study you need a sample that is representative of the whole area. So in this example you would need to ask yourself whether 15 pebbles is representative of the whole north or south end of the beach.

Essential notes

Formula for standard deviation

$$\sigma = \sqrt{\frac{\Sigma(x - \overline{x})^2}{n}}$$

Where:

σ = standard deviation

Σ = sum of

$\overline{x}$ = mean

n = number in the sample

Standard deviation is very useful when used to compare two data sets. In other words, you can use standard deviation when you want to compare the dispersion of two or more sets of data.

The standard deviation links the data set to **normal distribution**. In a normal distribution:

- 68% of the values lie within ±1 standard deviation of the mean
- 95% of the values lie within ±2 standard deviations of the mean
- 99% of the values lie within ±3 standard deviations of the mean

This is shown on the graph below:

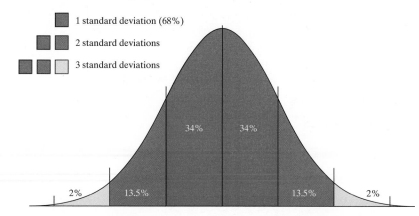

Fig 3
Standard deviation within the normal distribution curve

Now study the following example to show how standard deviation is calculated.

Calculating standard deviation

Refer back to the data on pebbles **(table 4)**. In the example below the calculation for standard deviation has been calculated for the northern end of the beach.

A	B	C	
Pebble size (mm)	$x - \bar{x}$	$(x - \bar{x})^2$	
58	31.7	1004.89	
43	16.7	278.89	
38	11.7	136.89	
33	6.7	44.89	
32	5.7	32.49	
25	−1.3	1.69	
24	−2.3	5.29	
24	−2.3	5.29	
23	−3.3	10.89	
19	−7.3	53.29	
19	−7.3	53.29	
19	−7.3	53.29	
14	−12.3	151.29	
12	−14.3	204.49	
11	−15.3	234.09	
$\sum x = 394$	$\bar{x} = 26.3$	$\sum(x - \bar{x})^2 = 2270.95$	$\dfrac{\sum(x - \bar{x})^2}{n} = \dfrac{2270.95}{15}$
			$\sigma = \sqrt{\dfrac{\sum(x - \bar{x})^2}{n}} = 12.30$

Table 6
Standard deviation calculation for the northern end of the beach

Step 1: Add up all the values of the pebble sizes in column A. In this case the sum is 394.

Step 2: Work out the mean: $\frac{394}{15} = 26.3$

Step 3: Subtract the mean from each value to complete column B.

Step 4: Next square each value in column B and insert in column C.

Step 5: Add up all the values in C. In this case the sum is 2270.95. Then divide this figure by the number in the sample: $\frac{2270.95}{15} = 151.40$

Step 6: Next find the square root of 151.40. The answer is 12.30.

Therefore, standard deviation = **12.30**

How do you use the standard deviation?

In the example given the mean is 26.3. This means that in a normal distribution graph 68% of the data should lie between 14.00 mm and 38.6 mm. These values are calculated by subtracting and adding 12.3 from/to 26.3.

Remember, the lower the standard deviation score, the more clustered the results are around the mean and the more reliable the mean is. For this figure to be of any use we now need to compare it to the standard deviation for the southern end of the beach.

Repeat the exercise for the results from the southern end and compare your result to the results below.

Standard deviation for the southern end:

Mean = 50.8

$$\frac{\Sigma(x - \bar{x})^2}{n} = \frac{7850.40}{15} = 523.36$$

The square root of 523.36 = 22.88

Therefore, standard deviation = **22.88**

This means that 68% of the data should lie between 27.9 mm and 73.7 mm.

Standard deviation suggests that there is more clustering around the mean at the northern end of the beach as the figure is smaller. The mean is therefore more reliable at the northern end. This is also supported by the other measures of central tendency which have indicated that there is less dispersion at the northern end of the beach.

A2 statistical tests: Mann-Whitney U test and chi-squared test

The Mann-Whitney U test

The Mann-Whitney U test is a technique that tests to see if there is a difference between the medians of two sets of data. It is **non-parametric**. This means it assumes that the data is not normally distributed. However, it does assume that there is similar dispersion of both sets of data (you can check this on a dispersion graph).

The test can be used if you want to investigate the differences between two sets of similar data. Some conditions apply:

- You can only compare two sets of data.
- The data must be ordinal (it can be ranked in order).
- You need a minimum of five values in each data set.
- It is not advisable to use more than 20 values in each data set as the exercise becomes unwieldy.

The Mann-Whitney U test starts with a null hypothesis:

There is no significant difference in the medians of the two sets of data.

Once you have calculated the value of U, you then have to compare it to the critical values. If the value of U is less than or equal to the critical value then

Examiners' notes

Both these tests are comparative tests that can support your hypotheses. Like Spearman's rank test, they are used to test the significance of your results. Tests of significance are used to tell us whether the differences between the two sets of sample data are truly significant or whether these differences could have occurred by chance.

you can reject the null hypothesis and accept that there is a difference in the two sets of data.

Mann-Whitney U test	
Strengths	**Weaknesses**
• You can use two data sets that have different sizes, e.g. one data set could have 10 values and the other only 8. • You can state whether the relationship is significant or if your results occurred by chance. • You can see clearly whether there is a difference in the median of two sets of data.	• It is a lengthy calculation and prone to human error. • It does not explain why the difference in the two data sets occurs.

Examiners' notes

In the exam you would develop the null hypothesis by linking the data sets, e.g. *There is no significant difference in hydraulic radius in the upper course and lower course of the river.*

Worked example of Mann-Whitney U test

In this example a student is investigating economic deprivation across the city of Plymouth. As part of her investigation she has used the National Statistics website (www.neighbourhood.statistics.gov.uk) to obtain secondary data on indices of deprivation. From her investigation she has formulated the hypothesis:

There is a greater income deprivation score for inner-city areas than outer suburbs.

She has found out the income deprivation score for eight **super output areas** (SOAs) (small census unit areas created by the National Statistics Office) in each of her two study areas – one in the inner city (St Peter and the Waterfront), and the other in the outer suburbs (Plymstock). The table below shows her results. The income deprivation score measures the percentage of people who are income-deprived. The higher the score, the more income-deprived the area is.

Income deprivation score	
Inner city (St Peter & the Waterfront)	**Outer suburb (Plymstock)**
0.53	0.05
0.40	0.09
0.24	0.06
0.25	0.12
0.08	0.05
0.12	0.08
0.16	0.10
0.20	0.13

Table 7
Income deprivation scores in two areas of Plymouth

The student decides to conduct the Mann-Whitney U test to see if there a difference between the two areas.

First she sets out the null hypothesis: *There is no difference in the income deprivation score between the inner-city and the outer suburb.*

She then follows these steps to complete the test:

Step 1: Label the data sets x and y. If they are of different sizes then label the smaller one x and the larger one y. In this case they are the same size so it doesn't matter which is which.

Step 2: Rank the scores in terms of their position in both samples – insert these in column B and column D. Where the scores are the same, take an average of the rank values.

NB Rank the lowest score as 1, second lowest 2 and so on.

Step 3: Total the ranks in column B and column D.

A	B	C	D
Inner-city scores (x)	**Rank (r_x)**	**Outer suburb scores (y)**	**Rank (r_y)**
0.53	16	0.05	1.5
0.40	15	0.09	6
0.24	13	0.06	3
0.25	14	0.12	8.5
0.08	4.5	0.05	1.5
0.12	8.5	0.08	4.5
0.16	11	0.10	7
0.20	12	0.13	10
	Total = 94		Total = 42

Table 8
Mann-Whitney U test rank table

Step 4: You now have to calculate the U values for both samples, using the formulae below:

$$U_x = n_x \times n_y + n_x \frac{(n_x + 1)}{2} - \Sigma r_x$$

$$U_y = n_x \times n_y + n_y \frac{(n_y + 1)}{2} - \Sigma r_y$$

where n = number in the sample. In this example it is 8 for both x and y.

The calculations for this example are thus:

$$U_x = (8 \times 8) + \frac{(8 \times 9)}{2} - 94 = \mathbf{6}$$

$$U_y = (8 \times 8) + \frac{(8 \times 9)}{2} - 42 = \mathbf{58}$$

Your result for $U_x + U_y$ should equal $n_x \times n_y$ – if they don't you have made a mistake!

Step 5: You now need to select the smaller figure of 6 and compare it to the table of critical values to test the significance of the result. If the value is lower than or equal to the critical value you reject the null hypothesis.

Sample size		Critical value at 0.05% significance level	
n_x	n_y		
8	8	13	

Step 6: You can now write a summary statement to express your result for the Mann-Whitney U test. In this case:

The lower value of U is 6. This is less than the critical value of 13 at the 0.05% significance level. Therefore, I can reject the null hypothesis and accept that there is a difference between the income deprivation scores for the inner-city area of St Peter & the Waterfront and the outer suburb of Plymstock.

It is clear just from looking at the values that there is difference in the income deprivation scores. The point of doing the Mann-Whitney U test is that it allows us to be certain about accepting or rejecting the null hypothesis.

Essential notes

A table of critical values for U can be found at: http://math.usask.ca/~laverty/S245/Tables/wmw.pdf

Chi-squared test

The **chi-squared test** (also referred to as the X^2 test) is used to investigate spatial distributions. It looks at frequencies or the distribution of data that you can put into categories, e.g. pebble shapes at different sites along a river's course or frequencies of plant types at different stages of a succession.

Chi-squared is a comparative test as it compares actual data collected against a theoretical random distribution of the data.

The data collected is called the **observed data**.

The theoretical, random distribution is called the **expected data**.

What is needed to use the chi-squared test?

- The data needs to be organized into categories.
- The data cannot be in the form of percentages and must be displayed as frequencies.
- The total amount of observed data must exceed 20.
- The expected data for each category must exceed 4.

As with other statistical tests, chi-squared requires and tests a null hypothesis. The null hypothesis is: *There is no significant difference between the observed distribution and the expected distribution.*

The strengths of chi-squared lie in the fact that, as with other statistical tests, you are checking the significance of your results. Chi-squared allows you to test categories and distribution.

As with other statistical tests, the weaknesses include human error in calculating X^2. It also doesn't explain why there is or isn't a pattern to the distribution. This will need further investigation.

Now work through the example below.

Worked example of chi-squared test (X^2)

A group of students investigated the orientation of pebbles in an exposed bed of glacial till. The glacial till was situated near the lip of a corrie in the Lake District. The students wanted to investigate whether there was a pattern to the orientation of the long axis of the till. Their hypothesis was: *There is a relationship between the orientation of the glacial till and the direction of the glacier.*

They measured the orientation of 40 pebbles and placed their results into four categories:

0–45° = 2 pebbles

46–90° = 10 pebbles

91–135° = 23 pebbles

136–180° = 5 pebbles

The data suggests that there is a preferential direction but as this could be due to chance a chi-squared test is carried out. The test begins with the assumption that there is no preference for any direction, with the null hypothesis:

There is no significant difference between the observed orientation of pebbles and the expected random orientation.

Next the students created a 'contingency table' shown below:

Orientation	Observed (O)	Expected (E)	A O – E	B (O – E)2	C (O – E)² / E
E					
0–45°	2	10	–8	64	6.4
46–90°	10	10	0	0	0
91–135°	23	10	13	169	16.9
136–180°	5	10	–5	25	2.5
					X^2 = 25.8

Table 9
Contingency table

They then carried out the chi-squared test, as follows.

Step 1: Enter the raw data into the Observed (O) column.

Step 2: Calculate the Expected (E) frequencies by adding up all the observed data (in this case 40) and dividing by the number of categories (in this case 4).

Step 3: Complete column A by subtracting the expected value from the observed value. Then complete column B by squaring O – E.

Step 4: Divide the figures in column B by the expected value (E) to complete column C.

Step 5: The chi-squared result (X^2) can now be calculated by totalling all of the values in column C.

Chi-squared value (X^2) = **25.8**

The result by itself is meaningless. You now need to test its significance.

Step 6: Work out the degrees of freedom using the formula (n – 1), where n is the number of observations – in this case the number of categories which contained observed data.

For this example, n = 4, so the **degrees of freedom** are 4 – 1 = 3

Step 7: Using the table (below, compare your X^2 result with the degrees of freedom for the 95% significance level. **If the X^2 result is the same or greater than the value given in the table, then the null hypothesis can be rejected.**

Degrees of freedom	Significance level	
	0.05%	0.01%
1	3.84	6.64
2	5.99	9.21
3	7.82	11.34
4	9.49	13.28
5	11.07	15.09
6	12.59	16.81
7	14.07	18.48
8	15.51	20.09
9	16.92	21.67
10	18.31	23.21
11	19.68	24.72
12	21.03	26.22
13	22.36	27.69
14	23.68	29.14
15	25.00	30.58

Table 10
Critical values of chi-squared

Step 8: Complete a summary statement:

At 3 degrees of freedom the X^2 result of 25.8 is above the 0.01% critical value of 11.34. Therefore we can reject the null hypothesis and accept that the orientation of the till did not occur by chance and is not randomly orientated.

Investigative skills and the assessment of fieldwork as part of AS Unit 2 and A2 Unit 4A

In both AS Unit 2 and A2 Unit 4A a range of geographical skills are assessed. Part of the assessments will always include questions on fieldwork undertaken during each course.

The whole of question 2 of AS Unit 2 is subdivided into a number of parts relating specifically to your own fieldwork and associated investigative skills. In order to prepare for these questions you will need to have completed a full investigation from the planning stage to final evaluation.

The investigative process at AS

This can be divided into a number of distinct steps.

1. Initial planning: including aims, choice of locality, research questions and hypotheses, followed by secondary research and data collection
2. Primary research and data collection: including sampling, safety considerations and risk assessment
3. Data presentation and analysis
4. Conclusions, evaluation and implications for further study

1. Initial planning

For many students the initial planning and choice of locality, along with at least the broad focus of their investigation, will have been decided by their teachers. However, there could still be questions linked to this early planning and so you need to be in a position to give clear answers.

The aim of an investigation is the broadest statement of intent; for example, '*to assess the changing downstream characteristics of a river*' or '*to consider issues relating to the opening of a new superstore*'. Alternatively, at this general level a simple statement could be made or research question posed, such as '*To what extent does distance from the shore affect the characteristics of sand dunes?*'

At a sharper level of focus one or more hypotheses may be set. These would state precisely what you are going to investigate and should be written in terms that can be proved or disproved, such as '*The width and depth of a river will increase downstream*'. This is where underpinning theory may need to be explained; for example, with reference being made to the Bradshaw model of downstream river development. This is also where you need to firm up your theoretical knowledge and do some secondary research on the location you are dealing with.

In choosing the location for your investigation, considerations might include: accessibility; safety; availability of resources, including data availability and manageability. However, in the end what you choose to investigate should be 'good geography' linked to the studies you have made in other parts of your course, and should allow you to come to some realistic conclusions.

Examiners' notes

You cannot afford to miss out on any of these stages since questions could be asked on any part of the investigative process. All aspects of the process will not be tested in any one paper but there could be different combinations appearing from time to time.

Essential notes

If you are designing your own investigation from the very beginning it may be worth carrying out a simple feasibility study to make sure that you do not waste your time and that the information you require is available to you. Ask your teacher for help here – it is not against exam regulations.

Examiners' notes

In the written exam you could be asked questions as simple as '*State the aim of your investigation*'; '*Describe the geographical theory or concept that formed the basis of your investigation*'; or '*Explain why the location of your fieldwork investigation was a suitable one*'. You need to prepare for simple short-answer questions such as these.

Case study: Hilbre Islands – an example of a fieldwork investigation

Location: Hilbre Islands, at the mouth of the Dee estuary off the west coast of the Wirral peninsula in northwest England. The northern edges of the islands face into Liverpool Bay and the Irish Sea. One side of the islands faces the Welsh coast, the other faces the Wirral coast. The islands are accessible by foot at low tide, but are separated from the mainland at high tide. Locate this area on an atlas.

Fig 1
Little Hilbre seen from Hilbre

Aims of investigation:

1. To study the effects of geological structure and lithology on coastal landforms
2. To investigate the processes acting on wave-cut platforms

Research statements or hypotheses

Research statement 1 (RS 1) (linked to Aim 1): *The detailed variation in rock characteristics and the rock structures have affected the detailed form of the cliffs and wave-cut platforms.*

Hypotheses 2 & 3 (HYP 2 & 3) (linked to Aim 2): *The size of rocks and pebbles (collectively called '**clasts**') will progressively decrease, and their degree of roundness will progressively increase, in moving away from the foot of the cliff across the wave-cut platform.*

RS 1 relates to the nature of these Triassic sandstones, which:

- Have very variable bedding characteristics and strength
- Are cut by numerous vertical joints
- Dip across the islands at about 10° North Eastwards.

The cliffs are vertical with classic erosional landforms and the surface shape of the wave-cut platforms varies from one side of Little Hilbre to the other. These aspects of the investigation contain both qualitative and quantitative research.

HYP 2 & 3 relate to the processes of sub-aerial weathering and mass movement and the daily tidal movement of waves over the wave-cut platform. Fresh falls from the cliff face would provide the largest, most angular fragments, which would become smaller and more rounded further out across the platform, where the swash and backwash of wave action would be in contact for longer each day. These aspects of the investigation largely comprise quantitative research.

Secondary research

Local Ordnance Survey maps at different scales were accessed, along with Geological Survey maps and Regional Guides. The Management Plan for Hilbre Islands was accessed through www.wirral.gov.uk, which in turn gave links to other productive websites such as Friends of Hilbre at www.deeestuary.co.uk. Aerial photographs and Google Earth links provided information on wave patterns.

Examiners' notes

You may be asked to write about safety considerations and risk assessment in the context of your own investigation. It may seem a straightforward question but you will need some specific detail to gain high marks. Nothing should be taken for granted in such answers and, as always, you should not leave the examiner to fill in the details for you.

Examiners' notes

Stratified and systematic sampling are often confused in exam answers. Make sure you know the difference. Be prepared to explain which you have used, and why. With **random** sampling students often forget to say how the random selection is obtained.

2. Primary research and data collection

Prior to any work 'in the field', safety needs to be considered. In many school or college situations a full **risk assessment** may be needed before any field trips can take place. Appropriate forms are usually available for this purpose. Any risk assessment first requires identification of actual or potential hazards and then an indication of how these can be overcome or reduced to an acceptable level. Some safety issues are very obvious, especially when working in exposed and remote physical environments, but it is also the slightly less obvious or less extreme events which need to be considered. For example, when working in a coastal location the actual risk of drowning is likely to be very small indeed, whereas slipping on rocks and twisting an ankle is much more likely. In an urban environment the dangers of traffic may be obvious but the need to think about how to carry out interviews and avoid the risk of being isolated, or receiving verbal or other abuse, may be important in your planning.

It is not usually possible to collect complete sets of data and so some form of sampling may be necessary. Which method you choose to use depends upon the nature of your investigation. The impression often given is that random sampling is usually best, since it should remove the risk of bias. This is not always the case in geographical investigations, since you will often be looking to recognize some sort of theoretical spatial distribution; this might suggest systematic or stratified sampling as more appropriate. In studying the downstream changes in a river there are advantages in having a systematic sample (equal spacing along the river) since you might want to demonstrate that downstream changes take place successively.

In applying a questionnaire linked to people's opinions about an issue you may need to give out questionnaires in proportion to the numbers of people potentially in each interest group, so you do not get skewed results on analysis. Remember, you can sometimes use more than one sampling technique at a time, for example if you want to have some structure in your sample but are concerned about potential bias. You might then decide to use some form of random sampling grafted on to a systematic or stratified approach.

Your data collection might involve the use of quite sophisticated equipment such as flow meters for river studies or callipers for measuring pebbles. Often you can improvise quite easily with floating objects, simple rulers, **ranging poles** and measuring tapes. Recognition charts are also readily available for stone roundness and plant identification. Questionnaires are almost always designed for purpose, as are interview questions. Data recording sheets are frequently needed, and although some of these are commercially available it is often best to design your own for your specific purposes. In the examination you may get the chance to explain how you collected your data, how you used your equipment, and how and why you designed your own collection and recording sheets.

Case study: risk assessment

A full risk assessment was carried out following local authority procedures. Particular issues relating to this coastal study location were:

- Access only at low tide – tide time planning needed
- Safe route out to islands to be followed
- Mobile phone available with emergency numbers known
- Contact made with countryside ranger on island
- Hard hats for work near cliffs, waterproof clothing and footwear
- Briefing and warnings related to work near cliffs and slippery nature of rock platform surfaces covered in silt and bladderwrack
- Students always working in pairs or threes
- Drinking water and sun cream available

The safest route to Hilbre Island

1. Start from Dee Lane Slipway, which is adjacent to the Marine Lake, West Kirby.
2. Walk toward Little Eye, the smallest of the three islands, keeping it on your right.
3. As soon as you pass Little Eye turn right and continue on the sand passing Little Hilbre (or Middle Eye) on your left.
4. Between Little Hilbre (also known as Middle Eye) and Hilbre take the rough track over the rocks towards the south end of Hilbre where there is a footpath leading onto the island.

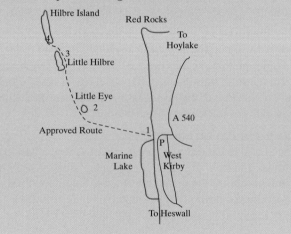

Fig 2
Diagram showing the safe route to Hilbre Island

Primary research and data collection, including sampling

For **RS 1** close observation, field sketching, taking photographs and making simple measurements of the cliff face were carried out using a small ruler, metre rule, ranging poles and simple **clinometer**.

Fig 3 (left)
Three-metre-high cliff on the Wirral side of Middle Hilbre with a natural arch and recently fallen blocks lying on a smooth wave-cut platform

Fig 4 (right)
Vertical seven-metre-high cliff on the Welsh side showing complex bedding planes in rocks which are dipping into the cliff face. The curved angled beds are 'current-bedded' sections. Vertical 'joints' allow rainwater and waves to penetrate

For parts of **RS 1** and **HYP 2 & 3** a data recording sheet was devised (**table 1**). This allowed the profiling of the wave-cut platform to be recorded, along with size of largest **clast**, median size of clastic material and mean degree of rounding for each measurement position across the wave-cut platform. Since the hypotheses anticipated that progressive changes would be identified in moving away from the base of the cliff, systematic sampling was chosen (every 2 m). Clasts were selected at each sampling point using a 'blind random' technique.

Table 1
Data collection sheet for wave-cut platform

Degree of roundness tally from 11 samples									
Distance from cliff (m)	Fall or rise of surface (+/- in cm)	Largest clast size (mm)	Median clast size (mm) from 11	Very angular	Angular	Sub -angular	Sub- rounded	Rounded	Well- rounded

Profiling the surface of the wave-cut platform was carried out using three ranging poles and a simple bubble level. This method is not highly accurate but it is enough to show the contrasts between the shape and slope of the wave-cut platform between the Welsh and Wirral sides, which was the purpose of this part of the investigation. It is also quick, which is necessary when working in the low tide 'time window'.

Fig 5
Students using simple equipment to survey the wave-cut platform

3. Data presentation and analysis

These go hand in hand, since the best way to present data is one that encourages or allows analysis to take place. Earlier sections of this guide have shown some of the many ways in which data can be accurately, meaningfully and attractively presented.

In presentation, bear in mind the following:

- In a geography investigation, methods of presenting material spatially (with maps) will be important.
- These may be based on existing maps or be specially drawn for the purpose in mind.
- Simple techniques often work very well, such as using overlays or using a map as a base on which to plot other information.

Examiners' notes

Examiners will be looking for detail and accuracy in your descriptions of what you did in your investigation. Don't leave anything out. State the obvious if necessary. The examiners will *not* do the work for you.

- Photographs, preferably well labelled, are almost always useful. They can also be used as the basis for well-constructed line diagrams or sketches.
- Computer graphics can help, and may be very attractive, but beware – you can use too many different kinds if you are not careful and you will never have to produce one in the exam.

Presentation should flow easily into analysis:

- Line and scattergraphs are often very powerful aids and when a trend line or a line of best fit is added they become analytical.
- If you are involved in an investigation which is based on strict 'hypothesis testing' principles don't forget the traditional and relatively straightforward techniques before you get stuck into deep statistical analysis. A balance between quantitative and qualitative approaches is often best.

Your presentation and analysis should run smoothly into your conclusions and evaluation.

Case study: data presentation and analysis

A study of the cliffs was presented using labelled photographs and sketches. The cliffs are highest on the Welsh side of the island, vertical and free from any vegetation cover. On the Wirral side of the island the cliffs are also vertical, or even overhanging, but lower, and have some deep cave development: a blowhole, and a cave, arch and stack sequence.

The different heights reflect the **dip** of the rocks, as shown in the sketches and photographs. The varied features along the Wirral side of the island are likely to be the result of pronounced sub-aerial weathering on the face of the cliff due to groundwater seeping along the bedding planes, plus the presence of a very soft 'marl' layer at the base of the caves in places. (Marl is a clay found in thin layers – 10 to 20 cm – within the sandstone.)

Essential notes

Statistics should be used with a purpose. Difficulty in drawing a line of best fit on a scattergraph may suggest to you that you should apply a Spearman's rank correlation test to see if there is a valid correlation between your data sets.

Essential notes

Before using statistical tests:

- Make sure that they are necessary, and that you have chosen the appropriate ones.
- Be sure you know how to interpret the results of the tests, taking into account the degrees of freedom for your data and confidence levels, as appropriate.

Most of your analysis will be in terms of written description and explanation. You need to be clear and precise in your expression, quoting your evidence and justifying your identification of trends, correlations and relationships.

Thinly bedded sandstone with vertical joints exploited by weathering and waves. Soft marl layer at base undercut by wave action

Plant roots penetrating bedding in cliff face – searching for the groundwater which emerges from dipping rock layers

Wave-cut platform slopes away from base of cliff at about 10° (angle of dip of rocks). Most surface debris washed away by wave action. Base of cliff visible along whole section

Staining where water emerges from cliff face

Fig 6
Section of cliffs on Wirral side of Middle Hilbre

Recently fallen angular 'joint blocks'. Foot of cliff obscured by debris

Little evidence of root penetration in dry cliff face

Vertical cliff face with irregular bedding in rocks. Lines of weakness along bedding planes and at vertical joints

Smoothing of rock surface by abrasion and rounding of pebbles by wave action at high tide

Fig 7
Section of cliffs on the Welsh side of Middle Hilbre

Profiling of the wave-cut platform showed a great contrast between the two sides of the island. Again, this can be explained by the dip of the rocks towards the Wirral. The action of the swash and backwash of the waves would be different because of the angle of the rock layers to the surface of the wave-cut platform. By staying on the island over a high tide the approaching waves can be observed as they pass over the rock platform and their contrasting actions noted and described.

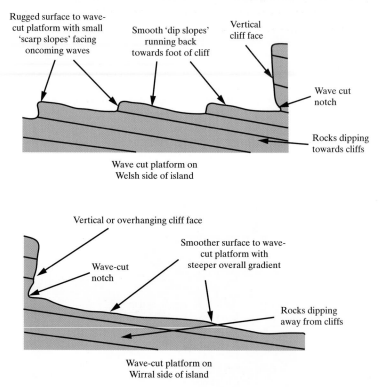

Rugged surface to wave-cut platform with small 'scarp slopes' facing oncoming waves

Smooth 'dip slopes' running back towards foot of cliff

Vertical cliff face

Wave cut notch

Rocks dipping towards cliffs

Wave cut platform on Welsh side of island

Vertical or overhanging cliff face

Wave-cut notch

Smoother surface to wave-cut platform with steeper overall gradient

Rocks dipping away from cliffs

Wave-cut platform on Wirral side of island

Fig 8
Wave-cut platform profiles

Scattergraphs were drawn for the largest size fragment, median size fragment and mean degree of roundness for a number of transects across the wave-cut platform. Starting positions for these transects were chosen from a random numbers table with different groups working on each transect and then sharing results. The scattergraphs showed variable results and lines of best fit were sometimes difficult to draw. The best correlations were for largest fragment size, and this was confirmed by using a Spearman's rank correlation test.

4. Conclusions, evaluation and implications for further study

The following steps and questions will help you to evaluate your fieldwork investigation:

- Look again at your initial aim or aims, commenting on the suitability of your chosen location for what you wanted to do.
- Review and evaluate your methods for collecting both primary and secondary data, pointing to strengths and weaknesses or limitations.
- Review and evaluate your choice of research question(s) or hypothesis(es), discussing their appropriateness in the light of what you accomplished.
- Develop your analysis into broader conclusions linked to textbook theory and/or what you found in your particular location.
- Were your conclusions to be expected, or was there something about your locality which threw up unexpected or unusual results?
- Would you do things differently if you were to start again?
- Could your work be developed or extended in some way into other investigations, such as if you move up to study at A2 level?

Case study: conclusions, evaluation and implications for further study

Conclusions for aspects of **RS 1** included the following:

- The character of the sandstone rock found on Little Hilbre does have a great influence on the detailed features of this section of cliff coast.
- The jointing, variable bedding and resistance to erosion of the rocks determine the shape and special features of the cliffs.
- There were a number of contrasts in the cliffs and associated features between the Welsh and Wirral sides of the island.
- The dip of the rocks determines the height of the cliffs but the shape and slope of the cliff face was not always what the textbooks suggest.
- The contrast in shape and slope of the wave-cut platform on either side of the island is related to the dip of the rocks.

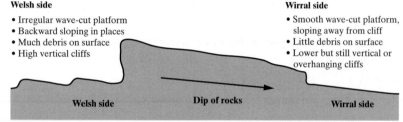

Welsh side
- Irregular wave-cut platform
- Backward sloping in places
- Much debris on surface
- High vertical cliffs

Wirral side
- Smooth wave-cut platform, sloping away from cliff
- Little debris on surface
- Lower but still vertical or overhanging cliffs

Welsh side   Dip of rocks   Wirral side

Fig 9
Sketch showing contrasts between the two sides of Middle Hilbre

Conclusions for aspects of **HYP 1 & 2** included the following:

- The variable results illustrated the fact that a coastal environment is complex and that clastic material can move towards or away from a cliff with the swash and backwash of the waves. Movement of material is more complex than in rivers.
- The need for multiple sampling to avoid bias was shown clearly, since specific transects produced different results and could lead to quite different conclusions being reached.

Some of the techniques for collecting primary data had their limitations and could be improved in future fieldwork. The most significant weakness relating to **HYP 2 & 3** was that fieldwork was carried out over a short time period. Visiting the location after different weather conditions, particularly in relation to wind (and so wave) strength, may well have produced contrasting results.

Further study could develop the idea of wave approach and the effects of this on the processes operating on the cliffs and the wave-cut platforms. These ideas will be picked up below when we turn to A2.

Some examples of AS fieldwork questions

1. Initial planning, including aims, choice of locality, research questions and hypotheses, followed by secondary research and data collection

State the aim of your fieldwork investigation and explain the geographical ideas that underpinned your work.

Clearly describe the location of your investigation and show why this was a suitable site for what you were attempting to do.

Outline the secondary research you carried out, showing how this contributed to the aims of your study.

2. Primary research and data collection, including sampling, safety considerations and risk assessment

Describe one method of primary data collection you used in your investigation.

Explain how you approached the issue of sampling primary data in your investigation.

How did you deal with the risks associated with your fieldwork investigation?

3. Data presentation and analysis

How did you use your ICT skills to help with your fieldwork investigation?

Outline one technique that you used to present one aspect of your primary research and explain why it was appropriate.

With reference to your fieldwork investigation, show you understand the difference between qualitative and quantitative data.

Describe and justify the use of one technique used to analyze your results.

4. Conclusions, evaluation and implication for further study

To what extent did your findings support the geographical ideas which underpinned your investigation?

*Evaluate the way you dealt with your investigation in terms of the collection of **either** your primary data **or** your secondary data.*

If you had the chance to extend your study further, what could you do to improve your understanding of the topic in question?

Examiners' notes

Carrying out some detailed
first-hand research on topics
such as these will give you
the chance to gather some
specific information and
extend your understanding
in preparation for the Geog
1 paper. Following the full
investigative process through
prepares you for question 2
on Geog 2.

Examiners' notes

In answering questions linked
to choice of location you
should be able to describe
concisely, but accurately,
your urban location and
the specific areas you
investigated. It is also likely
that you would have to give
some reasons for choice of
location(s). This is where you
can introduce aspects of both
your background theoretical
knowledge and more specific
knowledge of your chosen
location.

AS Human geography fieldwork

Fieldwork investigations linked to the core human section on population change in the AQA specification are well worth considering. The settlement case studies section requires a comparison of two (or more) of the following:

- An inner-city area
- A suburban area
- An area of rural/urban fringe
- An area of rural settlement

Reference should be made to characteristics such as:

- Housing
- Ethnicity
- Age structure
- Wealth and employment
- Provision of services

Investigating aspects of housing and environmental quality in an urban area

Location: Most urban areas will be able to provide appropriate material for a range of possible investigations. Some of the most productive areas for study are those which have suffered from economic decline in recent decades. Such examples include former mining and industrial towns and cities, ports and seaside holiday resorts.

Aims and hypotheses: A simple aim can often be directly linked to the AQA specification, for example: '*A comparison of an inner-city area with a suburban area in terms of housing*'. Equally, the focus could be on environmental quality, social deprivation or provision of services. Such aspects of urban geography are, inevitably, interconnected. A starting point for an investigation could be the useful summary deprivation diagrams obtainable on a **ward** basis from the **Office for National Statistics** (ONS) (available via www.neighbourhood.statistics.gov.uk). Insert a postcode under the Neighbourhood Summary section to show diagrams for particular areas (see **fig 10** and **11**).

Office for National Statistics

Neighbourhood   Statistics

Sign in or **Register>**

FIND STATISTICS FOR AN AREA<u>More about areas</u>

This search allows you to find **detailed statistics** within specific geographic areas, for example in **neighbourhood regeneration**.
First enter the name of an area OR full postcode:

e.g. Clapham Park e.g. N51JP
Then select the type of area you need statistics for:

- ☐   Local Authority

- ☐   Ward

- ☐   New Deal For Communities

- ☐   **Super Output Area**

<u>More areas</u> ▼
Search

NEIGHBOURHOOD SUMMARY

This search allows you to find a **summary report** for your **local neighbourhood**. If you want to know more about the neighbourhood you live or work in use this search.

Enter a full postcode from your neighbourhood:

e.g. N51JP

Search

Fig 10
Accessing **neighbourhood statistics**
via the ONS website: www.
neighbourhood.statistics.gov.uk

The Total Deprivation diagrams for two wards within Wirral Metropolitan Borough, reproduced below, show how good a starting point these can be. In this case the wards chosen represent a declining inner-city area bordering the river Mersey and a prosperous suburban area to the west bordering the Dee estuary.

The diagrams are based on a comparison of all wards throughout England and so show both local and national variations.

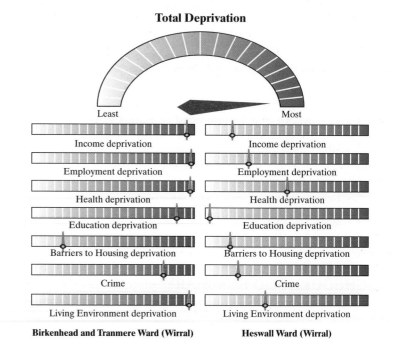

Fig 11
Neighbourhood statistics of deprivation for two wards in Wirral Metropolitan Borough – summary diagrams

This part of the ONS website also gives access to a very wide range of data which can be used in investigations. A specific hypothesis can be set up after considering the available data and perhaps some detailed study of local maps at different scales. A typical hypothesis might be: *'The quality of housing (or environment) in the suburban area will be higher than that in the inner-city area'*.

Safety considerations and risk assessment: It is important to consider safety when working in urban areas. Many of the risks will be different from, and perhaps less obvious than, those faced in a physical study in a remote and challenging environment. Apart from traffic dangers the hazards and risks are more likely to be linked to potential contacts with people, perhaps in an unfamiliar environment. If you are intending to interview people or carry out a questionnaire you need to plan carefully how you approach people and think about whether the questions might be provocative for some people. Mobile phones will probably be a part of your safety plan, but think about whether you should be openly and obviously using a phone in an unfamiliar place where you may be spending some time, rather than just passing through. Careful planning and a full risk assessment are clearly necessary.

Essential notes

Working in small groups is almost always necessary, whether you are approaching people for interview or not. Your teacher is the best guide when you are planning your primary data collection.

Primary research and data collection, including sampling: Inevitably, you will need to think about sampling in your planning. Random sampling may appear to be the most desirable to avoid bias but you may need to combine this with either systematic or stratified elements. For example, if you are looking at housing or general environmental quality within urban wards then you need to collect equal amounts of data from each ward to make comparisons valid. Similarly, you probably need to investigate different kinds of locations, such as points along main roads and others away from main roads, on housing estates or in areas of mixed development. Careful study of large-scale maps and preliminary visits will help you to design your sampling procedures.

There are a variety of environmental quality surveys available and these can easily be found via an internet search. A number of schools have put their own versions online. It is worth looking at some of these and then either adapting one for your purposes or designing your own from scratch (and so get something which is appropriate for your specific investigation). The focus can be on the quality of the housing itself, the appearance of the surrounding environment, commercial developments, traffic, nature and quality of local services, access to open space etc. Taking photographs throughout your survey is strongly advised. You can use these to present and analyze many aspects of your work.

Presentation and analysis: Data can be mapped for housing contrasts between and within two urban areas, for example dot or point mapping of house prices to show spatial differences. Dispersion graphing and box and whisker plots can be quite striking in showing differences between areas. Comparisons of means and medians can be made. Both house prices and environmental survey results can be presented and analyzed using different types of bar charts and histograms.

Well-labelled photographs provide a means both of illustrating and analyzing your findings (see **fig 12**, **13** and **14**).

Essential notes

In a study of housing quality and price, local free papers will give you much information, but visiting estate agents will add an extra dimension if you can interview a member of staff in the context of your study. Estate agents will be very familiar with different urban environments and how these affect house prices.

Some houses split into flats

Refuse bins on street in front of houses – no easy rear access

On-street parking – no garages or front drives

Houses in variable states of repair and maintenance

Fig 12
Inner-city environment, Tranmere

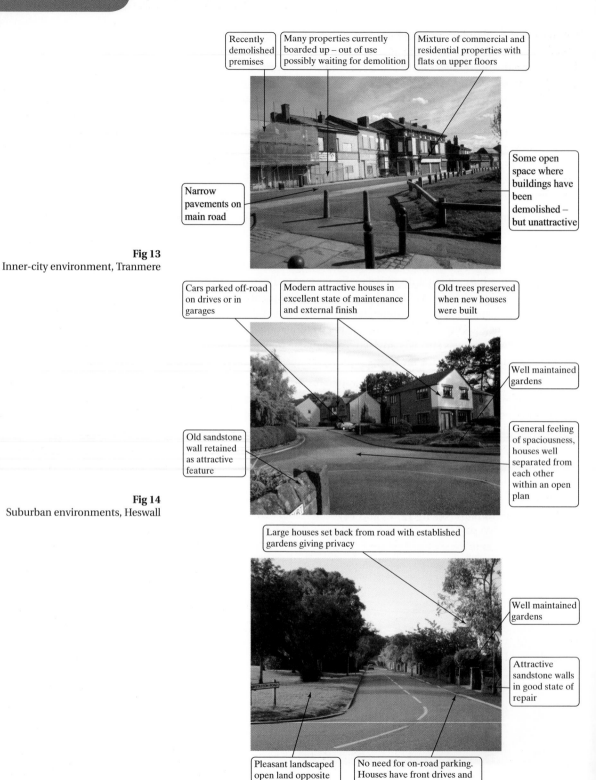

Recently demolished premises

Many properties currently boarded up – out of use possibly waiting for demolition

Mixture of commercial and residential properties with flats on upper floors

Some open space where buildings have been demolished – but unattractive

Narrow pavements on main road

Fig 13
Inner-city environment, Tranmere

Cars parked off-road on drives or in garages

Modern attractive houses in excellent state of maintenance and external finish

Old trees preserved when new houses were built

Well maintained gardens

Old sandstone wall retained as attractive feature

General feeling of spaciousness, houses well separated from each other within an open plan

Fig 14
Suburban environments, Heswall

Large houses set back from road with established gardens giving privacy

Well maintained gardens

Attractive sandstone walls in good state of repair

Pleasant landscaped open land opposite houses

No need for on-road parking. Houses have front drives and garages

Conclusions, evaluation and implications for further study: It is more than likely that studies of housing and environmental quality will show features which support the basic theory you will have encountered in your AS course. You will need to make such links clear in any examination answers; for example, how your investigation helps you to understand urban structure and morphology. Where you come across unexpected features, make the most of them. The real world is much more complex than the textbooks might suggest. Your particular location may have special attributes or a history of development which can go some way to explaining apparent anomalies. For example, your inner-city area may already have experienced some redevelopment and environmental improvement and your suburban area may include a small congested older settlement which has been engulfed by later growth.

Consider the issues below:

- Look again at your initial aim or aims. Were they successfully achieved? Review and evaluate your choice of research question(s) or hypothesis(es), discussing their appropriateness in the light of what you accomplished.
- Review and evaluate your methods for collecting both primary and secondary data. Did you have enough data to reach valid conclusions? Did you feel your sampling techniques were appropriate and valid?
- What would you do differently if you were to start your investigation again?

Finally, be prepared to make suggestions for further study. For example, having looked at housing and environmental quality you could look more closely at provision of services or aspects of crime within your study areas. There will be a wealth of statistical information available from the Office for National Statistics. Further primary research could be linked to people's perception of crime in their area or the level and accessibility of services. Such investigations carried out with the help of interviews and questionnaires could allow you to extend your work up to A2 level.

Essential notes

A good way of looking at the differences between the assessment of AS and A2 fieldwork is to think of AS as *what you did*, and A2 as *why you did it in that particular way*. This means that you will be evaluating your fieldwork, not just describing it.

Examiners' notes

These points may seem to be rather dry and theoretical but they contain the key ideas and many of the 'command words' you will have to face in the written examination. Examiners will set the questions based on statements such as these.

Essential notes

Keep your investigation small scale and clearly focused. Frequently candidates take on too much or try to cover too large an area. Another good idea is to pick an area which you can revisit if necessary. This is probably more significant for A2. If you are based a long way from your study area it may be difficult to return to check data, or to follow up a query or new idea.

Investigative skills and the assessment of fieldwork as part of A2 Unit 4A

The investigative process at A2

For section A of this examination paper you will need to have carried out a complete fieldwork investigation. This could be an extension of work done at AS level, or it could involve a completely new focus. However, the set questions will be more demanding than those at AS level since they have to demonstrate both 'synopticity' and 'stretch and challenge'. You will need to show a deeper understanding of the whole process of geographical investigation as well as the ability to handle more complex ideas and techniques.

The overall investigative process is the same as outlined for AS but some aspects need to be emphasized and extended. You will need to:

- Display an understanding of the purpose of the investigation and the relevant spatial and conceptual background
- Plan, construct and carry out sequences of enquiry
- Show an awareness of the suitability of the data collected and the methods used
- Be aware of alternatives and evaluate methodology
- Be familiar with alternative methods of data presentation/ processing
- Analyze, interpret and evaluate geographical information, issues and viewpoints and apply understanding
- Draw conclusions relating to the specific enquiry, and understand their validity, limitations and implications for the study
- Demonstrate an awareness of safety issues and risk assessment in geographical fieldwork
- Select and use a variety of methods, skills and techniques to investigate questions and issues, reach conclusions and communicate findings
- Use and understand your own experience of fieldwork and enquiry.

When choosing the focus for your fieldwork investigation you need to be sure that it does allow you to do all of these things. For this reason, investigations carried out at AS may not always be suitable to develop for A2. You will need to check this out in your initial planning.

There are no restrictions on the type of topic but it must:

- Be of a geographical nature – it is easy to wander into areas more appropriate to investigative work in other subjects
- Be linked to any content found in the AQA specification – this will help to keep the study relevant
- Involve primary data collection
- Be based on a small area of study – this is not defined precisely but the smaller the area studied the more likely you are to keep focused and find the investigation manageable.

Some examples of A2 fieldwork questions

1. Initial planning and decision making

State the aim(s) of your investigation and explain the reasons why you selected this aim.

Describe the location of your fieldwork investigation and explain why you felt this was relevant to the aims of your study.

Why did you choose your aim(s) or hypothesis(es)?

2. Methodologies

In the context of your chosen location, describe and justify the risk assessment you carried out.

Explain the significance of the sampling processes you used in your investigation.

Explain how one method of data collection you used was suitable for your investigation.

What improvements could you make to your data collection procedures to make your analysis easier and more meaningful?

3. Presentation

Explain why you chose the particular presentation methods you used. What alternatives might you have considered but chose not to use?

Evaluate the strengths and weaknesses of the data presentation methods you used.

4. Analysis and interpretation

Describe one data analysis technique you used and explain why this was suitable for your investigation.

Evaluate the relative strengths of the qualitative and quantitative techniques you used to analyze your data.

5. Conclusions and evaluations

How successful was your investigation in terms of your aim(s) or hypothesis(es)?

How was your geographical understanding further developed by carrying out your investigation?

Evaluate the usefulness of the conclusions you reached in your fieldwork study and point to any potential implications for further study.

To what extent did the findings of your investigation support or challenge accepted theory?

A2 case study: Physical geography – Hilbre Island

The coastal case study on Hilbre Island provided in the AS section suggests how an A2 fieldwork investigation might build on experiences during the AS course.

Initial planning and decision making

The findings of various aspects of the AS study showed that the processes operating and the landforms produced on and around Little Hilbre were not straightforward. The influence of geology was shown to be strong, but this did not necessarily explain all of the features seen or all of the contrasts noted between the Welsh and Wirral sides of the island. Some of the processes being experienced, both land/atmosphere-based (sub-aerial) and marine, may hold the key to some of these specific characteristics.

There is something special about the location which can be further investigated. However, prior to this, it is worth carrying out some extra secondary research to put the island group into its physical context.

Background geological and geomorphological summary from secondary research

The main elements of the geology of the study area are:

- Triassic sandstone rocks – part of the Bunter Pebble Beds (hundreds of metres thick) which outcrop in various parts of the Wirral and Merseyside. In the most recent publications these are now referred to as the Chester Pebble Beds.
- These are generally the most resistant of the local sandstones and so tend to stand out as prominent ridges in the region due to differential erosion (the stronger rocks form the upstanding ridges, and the weaker rocks form lower land in between the ridges because they are weathered and erode more easily).
- There is a very thin capping of glacial till (boulder clay) on top of the sandstone in places.
- The sandstone rocks of the islands themselves are very varied in nature but **dip** at an angle of about **8–10°** towards the Wirral shore and across the islands. This means that when standing in front of the cliffs the rocks appear as horizontal layers running along the face of the cliffs on both sides of the island.
- The sandstones of Hilbre are found in layers (beds) varying in thickness from a few centimetres to more than a metre. Some of these beds are very strong, some contain numerous large pebbles (hence the name 'pebble beds' for the whole sequence, although the layers with pebbles do not make up the majority of the rock layers found), and many of the beds show curved structures called **current bedding**. There is also a bed of marl (a very soft clay-like rock layer about 30–40cm thick) which is easily eroded. The differences in these rock layers mean that weathering and erosion take place at variable rates.
- The rocks of Hilbre are also faulted and jointed. The **joints** (natural vertical cracks in the rocks) are lines of weakness which sub-aerial

weathering and marine erosion both exploit. These are very common and can be seen in most sections of cliff face. **Faults** are much less common and are not easy to locate.

The main elements of the geomorphology of the study area are as follows:

- The whole area was covered in a thick layer of ice on more than one occasion during the Pleistocene ice age. This ice moved into the area from the Irish Sea basin and eroded the softer rocks which lay between the hard sandstone ridges of the Wirral. This emphasized the parallel ridges of harder sandstone, one of which formed what was probably one long island of Hilbre 12–15000 years ago. The ice also brought debris from as far north as Northern Ireland, southern Scotland and the Lake District. When these pebbles/boulders weather out of the overlying glacial till on Hilbre Islands they become part of the loose debris scattered over the wave-cut platform.
- Over the past few thousand years of the postglacial phase the islands have been split into three by marine erosion and rising sea level.
- The islands continue to be attacked by waves coming in from the Irish Sea (the longest fetch being across to Northern Ireland – over 200 km). Wave refraction around the North Wales and Wirral coasts further concentrates the main wave attack from the north west.
- Classic coastal landforms associated with cliff coasts can be found – cliffs, caves, stacks, wave-cut platforms and even a blowhole.
- The smallest island, Little Eye, is likely to disappear over the next 100 years or so, due to erosion, and particularly if sea levels rise with global warming.

The contrasts identified in the AS studies could provide some opportunities for further primary research. These are summarized below.

Little Hilbre contrasts between the Welsh and Wirral side		
	Welsh side	**Wirral side**
Cliffs	Higher (10–12 m)	Lower (3–5 m)
	Vertical, overhanging in places	Vertical, overhanging in places where caves have been formed
		Cave, arch, stacks found
	No arches or stack development	Cliff face partly vegetated, roots exposed in places
	No vegetation on cliff face	
		Fewer fallen blocks except where there are caves
	Many fallen blocks at base of cliff	
		Cliff face almost always wet with water oozing out from dipping bedding planes
	Cliff face usually dry (if not raining)	
		Cliff face dark, wet and often covered in algae and mould
	Cliff face clean and fresh	

Table 2
Comparison of Little Hilbre cliffs

Little Hilbre contrasts between the Welsh and Wirral side		
	Welsh side	**Wirral side**
Wave-cut platform	Very rugged	Smoother
	Slopes back to cliff in places	Slopes away from cliff at a steeper gradient
	Not much sand on surface, but fine silt covering on rock after calm weather	Pockets of sand over surface
	Rocks and boulders spread over surface, often trapped below small 'scarps'	Very few rocks and boulders on surface

Table 2 continued
Comparison of Little Hilbre wave-cut platforms

A number of interesting lines of further enquiry could come from these observations. Some of these can be linked to the geographical location of the islands.

Geographical location

This is worth investigating further. The position of the island group at the mouth of the Dee estuary, running in a line from southeast to northwest and pointing out into the Irish Sea, is significant in terms of the fetch of waves. This can be seen in the map below where it is clear that the only waves of any size are likely to come from the sector between west and north.

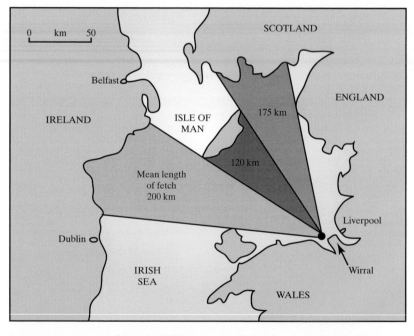

Fig 15
Extent and direction of fetch affecting Hilbre island group

Most waves approaching the Hilbre group of islands would inevitably be coming in at an angle to the coastline, further emphasized for westerly waves by refraction around the Welsh side of the estuary mouth. This is an opportunity to demonstrate the likely effects of some basic marine processes. The following aerial photograph (**fig 16**) shows the typical wave approach which runs along the islands rather than directly on to the cliffs on either the Welsh or the Wirral sides.

Fig 16
Waves approaching Little Hilbre from the Irish Sea – at a sharp angle on both sides of the island

Aims and research statements or hypotheses

With these ideas in mind a general aim can be defined in terms of '*investigating the effects of wave approach on the processes and landforms found around Little Hilbre*'. More specific hypotheses relating to the nature and movement of clastic material on the wave-cut platforms are also possible, such as:

- **Research statement A (RS A):** *The presence, distribution and movement of clastic material on the wave-cut platforms on either side of Little Hilbre will be influenced by the angle of approach of waves.*
- **Hypothesis B (HYP B):** *Material moved along the wave-cut platforms on either side of Little Hilbre will accumulate beyond the island and will show increasing roundness and decreasing size further away from the island due to abrasion and attrition.*

Collection of data

For research statement A: A number of transect profiles should be made across the wave-cut platform at right angles from the foot of the cliff on both sides of the island (starting points randomly or systematically chosen). These can be surveyed using three ranging poles and a bubble level, as suggested in the AS section. (Note: previous profiles produced for AS could be utilized.) Positions of any 'scarp faces' or other notable irregularities on the surface of the wave-cut platform should be clearly indicated and a recording made of the sections of rock platform which have any loose clastic material lying on the surface.

For hypothesis B: Ideally, three transect lines should be set out running southeast from the southern end of the island along the extended wave-cut platform which runs towards Little Eye. These would represent an extension of each side of the island plus a central line. This hypothesis gives an opportunity to gain experience of using the Mann-Whitney U test, knowledge of which is a requirement at A2 level but not at AS. A data collection point for this test should be located at the end of each transect line. At both ends of each line two sets of measurements should be taken and recorded:

1. For size, the long axis of a minimum of 15 pebbles should be measured and recorded
2. For roundness, measurements for the Cailleux index for each pebble, or a tally chart for Power's scale of roundness.

The Mann-Whitney U test can then be applied for each of the three transect lines. The test uses the medians which can be easily worked out.

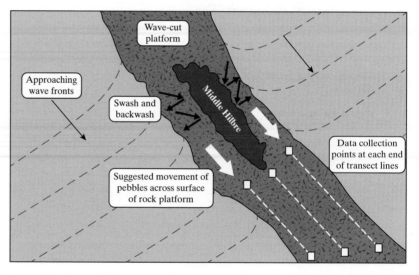

Fig 17
Suggested movement of material along wave-cut platform and locations for transect lines.

Presentation of data

For RS A: Accurate line profiles can be drawn with sections coloured, or a symbol added above the profile line, where loose clastic material was found. The different profiles can be superimposed one above another and the two sides of the island compared.

For HYP B: Simple dispersion diagrams can be drawn to give the visual contrast between the data sets for size and Cailleux index. If Power's scale was used for roundness then a divided bar diagram could be used to show the differences, if any.

Analysis

For RS A: Visual comparison between the profiles begins the analysis. Labelled diagrams (**Fig 18**) can then bring out the suggested explanations.

Welsh side
- Irregular wave-cut platform
- Backward sloping in places
- Much debris on surface
- High vertical cliffs

Wirral side
- Smooth wave-cut platform, sloping away from cliff
- Little debris on surface
- Lower but still vertical or overhanging cliffs

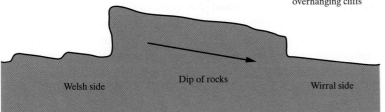

Fig 18
Contrasts between the Welsh and Wirral sides of Little Hilbre

For HYP B: Again, visual comparison between dispersion graphs begins analysis, followed by interpretation of the Mann-Whitney U test. This should indicate differences in the medians between the two ends of the transect line with smaller clastic material found further away from the island. The results for roundness are likely to be less convincing. These differences can be taken up in the analysis.

Conclusions and evaluation

General conclusions are likely to confirm three underlying ideas:

1. The influence of geology is still shown to be strong in the context of RS A.
2. The location of the island group and the islands' orientation in relation to the open sea is very significant in the movement of material over the wave-cut platform.
3. Processes acting in a coastal environment are complex.

Point 3, above, can lead to a general evaluation of the investigation in terms of limitations. The processes operating are weather-dependent and will vary from day to day and season to season. Unless monitoring of movement of material (e.g. by tracing painted pebbles) is carried out, many conclusions will be merely suggestions from incomplete evidence.

Examiner's notes

Make sure you really understand everything in the AIB. Find out, if you are in doubt. The topic could be linked to part of an optional unit you have not previously studied so use the time available before the examination to prepare yourself well. If the context is drawn from an optional area then material provided in the AIB will allow all candidates to be assessed to A2 standard

Examiner's notes

Don't expect that there will always be just one clear issue in the AIB. Often there may be a general topic with a number of connected but separate issues related to it. You can easily be thrown in the exam if you expect something to appear and then it does not.

Examiner's notes

Don't skimp on the preparation. Because you get the AIB before the exam itself, there may be a temptation to leave everything until you know what the topic is. This will damage your chances of getting high marks. Remember, this option is a complete A2 unit and needs adequate time spent in preparation. Practice in answering questions is essential.

Issue evaluation

This unit requires the use of the range of geographical skills, knowledge and understanding acquired during the AS and A2 course. This is done within the context of an issue evaluation exercise. An Advance Information Booklet (AIB) is provided on 1 April each year (for June examinations) and 1 November (for January examinations). Some further research on the issue, prior to the examination, may be suggested in the AIB.

What is an 'issue' in geography?

To most people an issue is a situation, conflict, problem or plan which invites discussion and usually produces a range of views, e.g.

- Plans for a new shopping centre away from the established town centre
- A bypass around a village
- Plans for a wind farm in an area of natural beauty
- Strategies for protecting a section of coastline from erosion
- Impact of a mining development within the developing world
- A flood-protection scheme

You cannot possibly prepare for every eventuality – and you do not need to. The AIB will provide you with the focus and will relate to elements within your AS and A2 courses. However, you do need to have been through at least one full issue evaluation exercise before embarking on the real thing. Such examples are available as past and specimen papers via the examination board website (www.aqa.org.uk), and other sources.

Preparation using the AIB

You must be familiar with the AIB before the examination. You will not see the questions beforehand but a thorough understanding of the material in the AIB will allow you to face them with confidence. You should:

- Familiarize yourself with the area represented in the AIB. This could be on a large regional, small local or whole country scale. Unless the AIB asks you to do more it would usually be enough to check on major physical and economic features along with political and population characteristics if you are working at country level. A general background knowledge would be enough in most situations, so you would be able to place any issues you identify in context. Working at a regional or country level, your starting point might be an atlas. Good atlases contain a wealth of information which would enable you to build up a solid background knowledge. Internet searches would give you access to a range of sites which could fill in some detail. However, the bulk of the information you need to answer the questions will be in the AIB.
- Read through the booklet a number of times to get the general feel of the topic. You do have enough time to become familiar with the material. You should be able to find detail within the appropriate sections of the AIB quickly under exam conditions. You will need to work efficiently and with confidence searching around the AIB.
- If there is anything you do not understand, to avoid losing time, do some research to find out. This need not necessarily be of a geographical nature. The materials will come from a range of sources and there may well be some unfamiliar terms used. With complex

tabulated data, make sure you understand how the categories used are defined. Questions will require a clear understanding not only of what they mean, but also of how they are relevant to the task(s). Weaker answers show confusion over the use of such data. Well-prepared candidates use detailed understanding to help them towards the highest marks.

- If the AIB suggests you do some specific extra investigation, do so. There may be websites listed or other suggestions for sources of information. There may also be references to possible fieldwork investigations which could be carried out in the area. Questions could then be set which relate directly to both of these areas of additional research.

The process of issue evaluation

The assessment itself will contain specific questions which address parts, but not the whole, of the issue evaluation process. This could involve a number of the following skills and processes:

- Prior additional research as suggested in the AIB
- Interpreting a range of data and other resources provided in the AIB
- Recognizing the shortcomings of the data and considering other possible sources through which these shortcomings might be overcome
- Use of techniques to present and analyze data from the AIB
- Considering how additional information could be collected using fieldwork, internet research and other methods
- Recognizing and defining the issue(s) in the AIB
- Relating the issue(s) to the body of geographical knowledge and understanding developed through your AS and A2 studies
- Considering evidence from different points of view
- Establishing criteria for evaluation of the issue or decision making
- Evaluating a range of options relating to the management of an issue or decision
- Identifying and analyzing potential areas of conflict and considering ways of resolving or reducing conflict
- Recommending how the issue(s) could be managed and possible impacts of such recommendations
- Review of the whole issue evaluation process

These final two aspects of the issue evaluation process are probably the most demanding and carry more marks than questions appearing earlier in the paper. Good answers will show a high level of critical analysis and are likely to demonstrate the complexity of the issues dealt with.

Before the examination it is often useful to tabulate ideas. This tabulation can be used for highlighting conflicts or for summarizing different **stakeholder** views. Columns could be headed advantages/disadvantages or for/against. More sophisticated versions might include a weighting based on quantitative data. For example, the issues over a proposed large wind turbine on a farm on the edge of green belt land might be presented as follows:

Examiner's notes

Ask someone to check your knowledge of the AIB. This could be in terms of some of the detail but also where the information comes from, both within the booklet and its original source.

Essential notes

There could be a geographical concept or theory which is new to you. Your teacher can help here but an internet search could also put you on the right lines. If there is a concept or theory that you have studied already in your geography course (remember this could be from your previous AS studies) then revise it.

Examiner's notes

In the exam you could be asked to use a technique for presenting or analyzing AIB material. Because of the limited time available such an exercise could not really be a very time-consuming one. An alternative might be to interpret and show your understanding of a graphical or statistical technique already applied in the AIB.

Examiner's notes

If you do use a tabulation or similar technique to summarize ideas or attempt analysis, do not reproduce this in the examination. These should be used at the preparation stage only. Examiners award marks for continuous prose answers where you can fully describe, explain and analyze. In the same way, avoid using bullet points in the exam – you will probably leave too much for the examiner to fill in for you.

Stakeholder	Views for	Views against	Implications for stakeholder
Planning officer	Follows general government policies for carbon reduction and expansion of wind energy production	Potential visual impact in area with strict planning controls Potential threat to visual amenity since location is close to a well-used picnic site	Need to balance conflicting local and national interests
Local residents	Landowner of proposed site in support?	Falls within attractive view over countryside from lounge and bedroom windows	Likely to affect resale value of house Could cause mental stress

Examiner's notes

Don't prepare detailed answers beforehand. You might just be lucky, but the risk of wandering off the point of the question, or missing it completely, is too great. Think about the questions which could possibly be set and prepare so that you are confident over a broad range of alternatives. Then expect the unexpected and you will not be disappointed.

Possible questions

It is always tempting to try to 'spot' questions based on the information you have in the AIB. This can be useful but it is potentially problematic. Examiners do not always set the obvious questions. For example, there may be some straightforward population density statistics tabulated for the wards of a borough, along with a blank outline map of the ward boundaries. Students may expect to be asked to draw a choropleth map based on this material. However, the exam paper may include a choropleth map already produced and the question may ask for a description of the distribution already shown on the map.

Sometimes students are so confident that a particular question will appear that they produce a readymade answer beforehand. The danger is that a question may appear which is similar but different in some critical way. The result is that the question set is not answered and the student scores few marks.

Think about the kinds of question which might appear in the exam but do not prepare in such a way that you depend on certain questions occurring. If you are confident about the content of the AIB, and have done any suggested further research, you should be able to handle almost any question which is set.

Past papers

Preparation using past papers and questions will always be worth doing well. You need to familiarize yourself with the range of different types of question which could be set in the examination. You also need to attempt at least one complete issue evaluation assessment paper in order to appreciate how a paper can be structured and how the different questions relate to each other. Sometimes the questions will lead up to a decision or an evaluation based on all that has gone before. In other papers this may not be the case – but all the questions will have some kind of link. For these reasons it might be wise to look at more than one specimen or past paper.

Examiner's notes

A full practice paper does not have to be attempted as a separate stand-alone exercise. There are opportunities during your main core or option studies when a complete paper could be integrated into your work. Discuss this possibility with your teacher.

Parties which may be involved in the issue

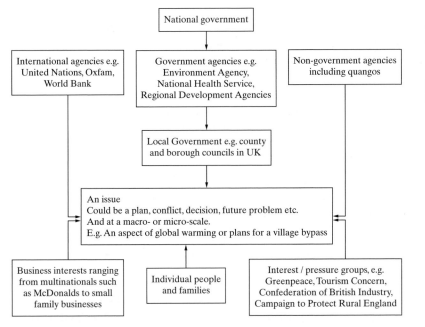

Essential notes

Quasi-autonomous non-governmental organizations (quangos) are set up by government for a particular purpose. Although they are often funded by government, they work independently.

Fig 1
Issue evaluation stakeholders. The stakeholders shown may overlap at times. The government-linked examples are from the UK. In other countries the way in which government bodies work will vary

It is likely that any issue you may have to deal with in the examination will involve considering the views of a range of groups, often referred to as 'stakeholders'. The specification for Unit 4B outlines the process of issue evaluation and parts of this process are likely to involve:

- considering the evidence from different points of view
- identifying and analyzing potential areas of conflict
- considering ways of resolving or reducing conflict
- establishing criteria for evaluation of the issue or for decision making
- evaluating a range of options concerning the management of an issue or a decision
- recommending a way of managing the issue or making a decision – and justifying the recommendation
- suggesting the possible impact of action that could result from the recommendation.

The specification also refers to the economics and politics of the process and the environmental context. All of the above are likely to be influenced by or have an influence on the stakeholders.

The links and considerations you will have to make are potentially complex and so you will need to understand, and perhaps do some further research on, the stakeholder positions. Understanding the stakeholder positions is fundamental to the issue evaluation process.

Examiners' notes

Because of the complex relationship between stakeholder positions you should be able to demonstrate high-level skills and understanding such as 'critical understanding'. This means that the highest marks are available for answers handling these ideas well.

A closer look at stakeholders

Government

This is probably the most complex broad group of stakeholders. Bear in mind that national and local governments act in different ways throughout the world. The points which follow relate to the structures and processes found in the UK.

Politicians make the laws and decide on policies but their civil servants carry out their wishes at a national level. This work is carried out through departments such as Health, Environment, Food and Rural Affairs (DEFRA), and Transport. Many of these departments have government agencies which work for and report to them such as the Water Services Regulation Authority (Ofwat) and the Forestry Commission working on behalf of DEFRA. You do not need to have a detailed knowledge of these complex, and frequently changing systems. However, if the examination AIB contains references to, and information provided by, any of these bodies then it makes sense to find out how they fit into the overall structure of government. In the UK, national governments often carry out their policies at a regional level because some regions have their own particular needs and priorities, for example in the context of regeneration of old industrial regions. For many aspects of planning and delivery of services it is the local authority, such as the county or borough council, which makes the decisions and carries out new developments, guided by overall government policies. Most planning decisions about changing land uses are in the hands of local planning authorities. Conflicts which could appear in an issue evaluation examination are often centred on local planning decisions.

At a larger scale the European Union operates through a number of departments which concentrate on particular policy areas, such as the Community Fisheries Control Agency (CFCA).

Non-governmental agencies

In recent years many non-governmental organizations (NGOs) have been set up to help develop aspects of government policy but within an independent framework. They can be small locally based groups or they can operate at a national or international level.

International agencies

Many international agencies are well known and are often focused on the environment or human rights and poverty issues. The United Nations (UN) is an organization which has many sub-groups such as the Intergovernmental Panel on Climate Change (IPCC), which was set up by the United Nations Environment Programme (UNEP) and the World Meteorological Organization (WMO). Another UN agency is United Nations Development Programme (UNDP), which is a significant body concerned with the developing world and the setting of the Millennium Goals which you may have come across in your geographical studies. Many

Examiners' notes

In the examination avoid referring generally to 'the government'. It is almost always the case that you can be more specific than this. Also avoid the common error of confusing local and national government. Weaker students often do this.

international agencies are charitable ones such as Oxfam and Médecins Sans Frontières. Others are essentially economic such as the World Bank and the World Trade Organization (WTO).

Interest/pressure groups

Pressure groups can also be international, such as Greenpeace, or they can be local and small scale, concentrating on a specific local issue. Such groups may take action in demonstrations, which often make media headlines, but mostly they work by lobbying politicians and public bodies in order to get their voices heard.

Business interests

Organizations within this group of stakeholders are often very powerful, with strong economic and political lobbying powers. Even individually large corporations carry much weight because of their potential contribution to economic growth and development in both a national and local context. General business interests are looked after by organizations such as the Confederation of British Industry (CBI), and at a global scale by groups such as the International Business Leaders Forum (IBLF), which was set up by the Prince of Wales, and major international companies such as BP, Volkswagen and Coca-Cola. The IBLF has a particular aim to promote responsible business, particularly in emerging economies. The stereotypical view of business interests may be that they are always selfish and greedy, and always looking to make maximum profits. Whatever the characteristics of a particular business issue the fact is that businesses cannot afford to act without taking into account political, environmental and social conditions within the context of their operations.

Individual people and families

Individuals can be involved in issue evaluation scenarios. Very often the **NIMBY** mentality can be suggested where people raise objections to specific developments. However, this is where you have to be careful not to simply write off a possible view that may be held. At the very least you should attempt to explain why a particular view is held, such as the possible effect on the market value of a home, blocking of a view or a noise nuisance. Often individuals group together to form a temporary pressure group if they see a common threat linked to a planned development such as a new road or potential loss of an environmental amenity such as a footpath or area of greenbelt land.

Not every issue evaluation scenario would include all of the above stakeholders. Sometimes there may be only one, although this would probably be quite unusual.

Examiners' notes

Be careful with stereotypes. In many cases a stereotype may be appropriate but be prepared to dig a little deeper and at least justify why a stereotypical view might be held. That's the way to get to the top marks.

Economics and politics in the issue evaluation process

Most issues have an economic aspect to them and for most stakeholders there is likely to be an economic dimension which is of concern to them:

- For business organizations the issue in question might provide an opportunity to expand or create new business in an area. Such opportunities can range from a large multinational company wishing to set up operations such as a mining/industrial development, or a holiday resort complex in the developing world, to a small local business looking to develop an activity centre in a UK national park.

- Businesses look to make profits for the company and, if it is a plc, for the shareholders. They have to balance costs against benefits when making decisions about whether to go ahead with investment schemes. In your preparation before the exam it can sometimes be useful to consider issues from a **cost-benefit** point of view. This kind of **analysis** can also be carried out for other stakeholders from the public or voluntary sectors. Again you can summarize ideas in a tabulated format listing costs and benefits and possibly ranking ideas with the most significant listed first. Remember, this should only be done as part of your preparation, not in the exam itself.

- The general economic climate may need to be taken into account. The global banking crisis from 2008 onwards, and subsequent worldwide economic recession, have had significant effects on investment and business confidence.

- Costs and benefits are not only concerned with capital investment and profits. There may be significant social costs and benefits in any issue under consideration. These social considerations are often of great importance within the **public** and **voluntary sectors**. It is often these considerations which are particularly complex and subtle. There are few absolute 'right or wrong' views. How you handle these issues is often key to gaining the top marks.

- It might be worth thinking about the idea of 'opportunity costs'. In simple terms this is the cost of something in terms of an opportunity lost. For example, if a town decides to locate a new campsite on a plot of vacant land it owns, the opportunity cost is whatever else might have been done with the land and money spent – such as to allow a factory to be built which might provide more jobs.

- It is difficult to separate the economic and political aspects of an issue and it is in this context that many social considerations are to be seen. Thus a government, at national or local level, may need to make decisions on spending taxpayers' money in order to gain certain social benefits. In recent years in the UK the broad policy of 'social inclusion' has been a key consideration in many decisions. Social benefits have often been 'weighted' in this way by local and national politicians, for example when investment has been needed for job creation or provision of facilities for the underprivileged. However, the political climate can change and with it such broad policies may carry less weight or be abandoned completely.

Examiners' notes

Remember that exams are set well in advance of when they are taken by students. By the time the AIB reaches you some material may already be out of date. New local, national or global conditions may exist. In such situations some extra research may be needed to bring aspects of the AIB up to date.

Examiners' notes

If you refer to ideas such as 'cost-benefit analysis' or 'opportunity costs' you will not get credit for simply using the terms. What such ideas will provide is a way of looking at issues. They should encourage you to demonstrate the complexity of issues and how there are interrelated elements within any issue. This is how to get the top marks.

Environmental context

The specification for the issue evaluation option encourages candidates to look at the following in an environmental context:

- The impact of the issues on the environment
- Conservation and exploitation
- Sustainability and growth

The issues you will be able to identify, and have to deal with in the examination, will almost always have an environmental aspect to them. This can be in a rural (remote or accessible) or urban (local scale or whole-town scale; industrial, inner-city, suburban etc.) context. In all cases the environmental considerations may be a key element in the main issues investigated.

Exploitation of physical resources has in the past had enormous negative impacts on the environment. In the UK and other developed countries environmental controls and restrictions are usually in place to protect and conserve sensitive environments such as wildlife habitats and areas of scenic beauty. In less developed areas of the world this may not be the case. The desire to exploit physical resources, such as raw materials for industry, or to set up new industrial production units, often based on cheap labour, can mean that the environmental consequences are virtually ignored. This is where the political as well as economic situation needs to be appreciated in considering issues arising in such locations.

The concept of sustainability is frequently raised in geography, and particularly in the context of the environment. The generally accepted definition of sustainable development (derived from the United Nations Brundtland Report) is 'development that meets the needs of the present without compromising the ability of future generations to meet their own needs'. If you don't find this definition helpful, consider the following four characteristics of sustainability:

- Reducing negative impacts
- Increasing positive impacts
- Recognizing both natural and human environments
- Looking towards the future

If you can clearly show the relationship between each of these aspects when considering sustainability and growth you will be able to demonstrate high level issue evaluation skills. A good example of where such skills were required was in the Issue Evaluation paper of June 2010. In the context of the North West Highlands Geopark, questions were asked on:

- The idea of a thriving rural community
- Issues that the physical geography of the area presents for settlement and economic activity
- The extent to which initiatives such as the creation of the North West Highlands Geopark can help rural communities to survive as viable communities in the 21st century

Examiners' notes

In your initial preparation, after receiving the AIB, the political background of the location should be considered. This may involve a little further research, particularly if the location is outside the UK.

Examiners' notes

Remember, impacts on the environment are not always negative. Students often tend to stress the negative aspects, emphasizing actual or potential conflicts. However, once environmental issues have been raised they are rarely ignored, and positive impacts often follow. Make sure you recognize the positives where they exist.

Examiners' notes

When considering environmental impacts, if you can appreciate and show how the political and economic situation has an effect on what can or cannot be done you will be more likely to demonstrate high levels of critical understanding and evaluative skills. High marks are awarded for such evidence.

Examiners' notes

In your answers, don't just throw in terms such as 'sustainability': unpack them to show exactly what you mean and why these ideas are significant.

Below you will find sample questions followed by sample answers, and also (tinted boxes) containing the comments and advice of examiners.

Unit 2, Geog 2, Section 1: Cartographical, graphical and statistical skills

In the Geog 2 exam there are two sections. The first section is based on either of the core sections from Unit 1 and tests geographical skills. The second section is based on students' own fieldwork and investigative research skills.

Make sure you are familiar with the question types. Past question papers and mark schemes are available to download on the AQA website, at: www.aqa.org.uk.

*1 (a) Study **fig 1** below, a divergent bar graph showing the monthly rainfall variation compared to the monthly norm in Keswick, 2009.*

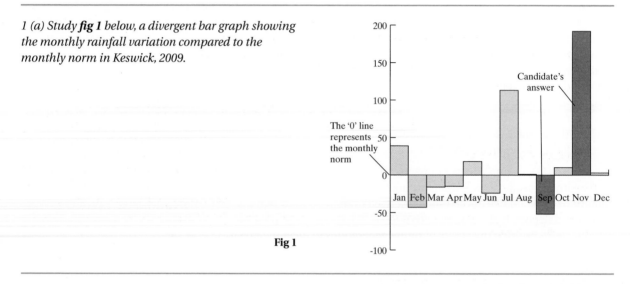

Fig 1

1(a) (i) Complete the graph by adding the following information. [**2 marks**]

Month	Monthly rainfall variation (mm)
September	−52
November	192

Strong answer
(The candidate has shaded in the two bars on figure 1, above.)

It is really important that you don't miss this type of question out. Many candidates fail to complete the graph – often because they haven't read, or even seen, the question. Also, if you leave this blank you may interpret the graph incorrectly and lose marks in subsequent related questions. ☞

The candidate has fully completed this task. The bars have been placed accurately and the shading (i.e. block) matches the other bars. Therefore full marks are awarded.

1(a) (ii) *Describe the variations in monthly rainfall shown in **fig 1**.* [**4 marks**]

Note, the command word here is 'describe'. Therefore do not attempt to explain the graph – you will not get any marks for doing so. When describing any graph, always look for trends and patterns first. Make sure you refer to actual values – there is often a mark for manipulating data.

Average answer

The rainfall varies from month to month for example in January there was 40 mm and in May there was 15 mm. The highest amount was in November and the lowest was in September. There are more months with rainfall above 0 than below. This is because it is very wet in Keswick and floods a lot. There is more rain in Keswick because it is a mountainous area and gets a lot of relief rainfall and also it is in the west of the UK. This causes the town to flood regularly.

This answer is point marked and would score 2 marks. There is an attempt to describe the pattern but the candidate has not fully understood the graph and appears to have interpreted it by referring to actual rainfall totals. Much of the answer is irrelevant as it is attempting to explain the graph. Valuable time has been wasted here by the candidate not answering the question.

Strong answer

You can see that there is considerable variation in rainfall in Keswick. The amounts fluctuate above and below the mean. The highest rainfall above the mean is November (192) and the least is September (−52) with a range of 244. There is no clear pattern but more months had above average rainfall than below average. Several months, such as August and December, were close to the mean.

This is a well-constructed response. It is fairly simplistic but the candidate has scored the full 4 marks. There is an attempt to look at overall pattern and clear reference to actual values. There has also been some manipulation of data which has been given credit. Note that there is unlikely to be more than 1 mark for a manipulation activity so do not waste time doing more than one calculation.

1(b)

Table 1 below shows the monthly rainfall amounts (mm) in Keswick for 2008 and 2009. For each year the mean has been calculated.

Year	Jan	Feb	Mar	Apr	May	Jun	Jul	Aug	Sept	Oct	Nov	Dec	Mean
2008	189	57	96	63	23	73	98	106	111	128	58	85	90.6
2009	115	17	51	31	71	38	182	76	24	90	269	87	87.6

Table 1

*The standard deviation has been calculated for 2008 and is shown in **fig 2** below.*

Standard deviation formulas for 2008 and 2009	
2008	**2009**
Standard deviation formula	**Standard deviation formula**
$\sigma = \sqrt{\dfrac{\Sigma\,(x\text{-}x)^2}{n}}$	$\sigma = \sqrt{\dfrac{\Sigma\,(x\text{-}x)^2}{n}}$
$\sigma = \sqrt{\dfrac{19682.9}{12}}$	$\sigma = \sqrt{\dfrac{58266.2}{n}}$
where: σ = standard deviation v = number in the sample	where: σ = standard deviation v = number in the sample
Standard deviation for 2008	**BOX A: Standard deviation for 2009**
σ = 40.5	σ = 69.7 (*This is the candidate's answer*)

1b (i) Using the formula for standard deviation shown below, calculate the standard deviation for 2009. Part of the calculation has already been done for you. Place your answer in the box marked 'A'. [1 mark]

The candidate has worked out standard deviation correctly and therefore gained 1 mark. This is a simple calculation providing you understand the formula. All that was required was to divide 58 266.2 by the number of months (12) and then find the square root of this. This can be done very easily on a calculator, which is why it is so important to remember to take one into the exam.

Fig 2

1(b) (ii) Using your result and the 2008 figure from 1(b) (i), write a statement to describe what the values for standard deviation tell us about rainfall in 2008 and 2009. [**2 marks**]

Strong answer
The standard deviation score is lower for 2008 meaning that rainfall is more clustered around the mean. In 2009 the figure is higher meaning rainfall is more variable.

This candidate clearly understands standard deviation and is able to describe what the results mean. This gains the full 2 marks. Note that the candidate gained the full 2 marks in the first sentence.

*1(c) Using map evidence, compare the valley and channel characteristics in grid squares 2924 and 2926 on the map extract of Keswick (Cartographical skills, **fig 1**, p 5).* [**6 marks**]

If the question asks for map evidence, you must give this in the form of grid references, place names or, if appropriate, height. The word 'compare' means you must look for differences and not deal with both grid squares in isolation. The question also requires you to compare two characteristics: valley and channel. To get full marks you must ensure that you cover both of these.

Average answer

Grid square 2926 shows that the channel is narrow with a V-shaped valley in the upper course. The channel will be deep with large boulders. The valley is also very narrow and there isn't a floodplain. The valley had very steep sides as the contour lines are close together. Grid square 2924 is in the middle course of the river. The channel is quite wide and it has some meanders near Brundholme woods with river cliffs and slip-off slopes. There is also a wide floodplain.

This answer is restricted to level 1 as the candidate has described the valley and channel characteristics of both grid squares but has not compared them. Both grid squares have been described in isolation. The reference to the depth of the channel is not creditworthy as it cannot be seen from the map. However, there is some good use of map evidence, such as reference to place names, which results in the answer reaching the top of the level and scoring 4 marks.

Strong answer

In grid square 2926 the river channel is much narrower than the river in 2924. The river is a small stream in 2926 but it is about 25 m wide in 2924. In 2924 there is also evidence of meandering (295246) whereas there isn't in 2926. The valley is v-shaped in 2924 with very steep sides on both sides of the river as the contour lines are close together. The valley in 2924 has a wider bottom, forming a floodplain. South of the river Greta the valley slopes more gently than the northern sides where it is steeper in Brundholme Woods.

This is a well-constructed answer that has covered all parts of the question. The candidate has clearly compared the two grid squares and has used map evidence to support the answer. The use of map evidence makes it clear that the candidate has done this part of the answer very well. The question is level marked and easily accesses level 2. This answer would gain the full 6 marks.

1(d) *Fig 3 is a picture of the Glenderaterra beck, a tributary of the river Greta. Label the picture to show characteristics of the river channel.* [**4 marks**]

Labelling a photo is a basic skill but there are some key points:

- The label must connect with the feature that is being labelled.
- Don't write one or two words, clearly describe the feature using geographical terminology.
- The feature must be seen in the image – don't label things that aren't visible, e.g. sounds or things you think should be there.

Large, ✓ angular ✓ bedload that hasn't had time to erode

Deep river channel because of vertical erosion

Narrow river channel ✓ – vertical erosion is taking place

V-shaped valley

Fig 3

Strong answer
(See the labels on the image.)

There is a variety in the quality of the labelling. The ticks show where credit was awarded. Note that 2 marks are awarded for the top left box as there are two separate characteristics – the size and the shape of bedload. The labelling of the narrow rivers channel ☞

is very good as the candidate has drawn a line to show the width of the channel. The two labels on the right do not get credit. The top one is not creditworthy as you cannot see depth of the channel in this image. The lower label is not relevant to the question as this is a valley characteristic not a channel one. The answer scores 3 marks.

1(e) *Study the OS map detail of Keswick (Cartographical skills, **fig 1**, p 5).*

*There was severe flooding in Keswick in November 2009. Using the Keswick map, and the divergent bar graph (**fig 1**, p. 76) explain the possible causes for flooding in Keswick.* **[6 marks]**

Make sure you refer to both the map and the bar graph. This is not a question about general causes of flooding – the response must be specific to Keswick. Note the question asks for *causes* – therefore you must give at least two different causes for full marks. Note the command word is 'explain' – you need to identify the causes and give reasons why it causes flooding in Keswick.

Average answer
The map shows that Keswick is built on a large meander; this means that the river will slow down when it flows through Keswick. Also the area of Keswick is built up so there is less infiltration due to the tarmac and concrete. This means that surface run-off is quicker and water reaches the river more quickly reducing lag time. Keswick is surrounded by large hills so there will be fast surface run-off and more rainfall. ☞

This answer is level marked and would only access level 1 because there is no reference to the graph. References to the map are very good – there are two different causes identified and they are both explained. The second cause (being built up) is well developed with several linking points. The answer would gain 4 marks at the top of level 1. This is a good example of why it is so important to make sure you answer all parts of the question.

Strong answer
The map shows that Keswick must be very prone to flooding as it is built on a large meander, meaning that water will flow more slowly around the meander bend. The town itself is built up with many roads such as the A591 and this means that there will be less infiltration on the concrete and tarmac, meaning that water will reach the rivers much more quickly. There will also be a network of drains and these also allow water to reach the river more quickly reducing lag time. ☞

The graph shows that rainfall varies considerably and some months have a much greater than average rainfall for example November. In months like this the ground will become waterlogged and increase surface run-off. Rivers will reach their capacity and not be able to cope with any more water and will therefore flood.

This answer accesses level 2 and achieves the full 6 marks. All parts of the question have been answered – there is reference to both the map and the graph. More than one cause of flooding has been explained.

Unit 2, Geog 2, Section 2: Fieldwork investigations: physical

This section contains students' responses and examiners' comments for a typical AS Section 2, question from Geog 2. The answers relate to fieldwork carried out by students in the context of the Hilbre Islands case study.

Part (a)

Clearly describe the location of your investigation and show why this was a suitable site for what you were attempting to do. [**4 marks**]

Average response

I chose Hilbre Islands for my investigation. There are three small islands arranged in a line close to the mouth of the River Dee between the Wirral and the coast of North Wales. I worked on Middle Hilbre which is about 200 metres long. The islands can be walked to at low tide.

This student gives a clear description of location with an indication of scale. However, there is no attempt to answer the second part of the question. 2 marks, the maximum allowed just for location.

Strong answer

The location I chose for my fieldwork was Hilbre Islands which are situated about 1 km off the coast at West Kirby at the northwestern end of the Wirral peninsula. The three islands, Hilbre itself, Little Hilbre and Little Eye are about 2 km long altogether. They lie in a line parallel to the coast at the mouth of the estuary of the River Dee. They point out into Liverpool Bay, which is part of the Irish Sea. This is a good site for me since I live in Hoylake close by and so I can visit the islands easily and revisit if I need to check on something or collect more data. I had seen photographs and maps of the islands and knew that there were sandstone cliffs, a stack and a rock platform all around the islands.

The question is point marked - 1 mark per point but with the possibility of gaining extra marks for any developed idea. This student gives a comprehensive description of location with an indication of scale. This would gain 2 of the 4 marks. The idea of convenient accessibility and the recognition that return visits might be needed gains extra credit up to the maximum 4 marks. The question has been fully answered in relation to clearly describing location and showing why it was suitable. 4 marks.

Part (b) (i)
Outline the secondary research you carried out,
showing how this contributed to the aims of your study.
[4 marks]

Average response
The local Wirral council website had links to a useful management plan for the islands and information sheets which gave me some geological background information. This was useful since I wanted to show how the geology of the islands had an effect on the landforms found there.

> The question is point marked.
> One source of secondary information has been identified and a simple attempt made to link this to the study aims. 2 marks.

Strong answer
Most of my secondary research was done on the internet following a Google search on Hilbre Islands. One of the best sites was the local Wirral council site which also gave me links to other useful sites such as 'Friends of Hilbre'. Via the council site I got information sheets and maps at different scales and a link to a webcam site so I could see what was happening on the islands. I visited the local planning office to get a very detailed large-scale map of the islands but was told that it had been produced more than 30 years ago and may not be completely accurate because of erosion and deposition which would have taken place since then.

> Some appropriate internet research had been carried out and some good material found from appropriate sources. Credit can also be given for the non-internet research. The answer falls down on the lack of a clear link to the aims – which is a common failing in this type of question. 3 marks out of the 4 would be awarded.

Part (b) (ii)
Describe one method of primary data collection you used in your investigation. [**6 marks**]

Average response
We used ranging poles and a level to survey the wave-cut platform. These are two metres long and were painted red and white. Groups of three had to work together because it was difficult to hold the equipment and record our findings on the recording sheet. The horizontal pole had to be kept perfectly level. We worked from the foot of the cliff out to the edge of the wave-cut platform, recording the changes in level.

This question is level marked.
The answer does not get into level 2. It refers simplistically to the equipment and to problems of using it. The answer lacks detail and there is no reference to sampling for choice of survey site. 2 marks.

Strong answer
I used three ranging poles, a metre rule and a small bubble level to survey the surface shape of the wave-cut platform on Little Hilbre. Three people worked together to hold the poles and the level and take down the rise or fall measurements on a recording sheet. The starting point for each profile was chosen from a table of random numbers to avoid bias if we just picked a nice-looking section of platform. Along the survey line we worked out the rise or fall every two metres, which was the length of the poles. We did this by holding two of the poles vertical and placing the other horizontally (checked by the bubble level) between these poles. The rise or fall could then be measured using the metre rule. If we came across a sudden rise or fall in the middle of a two-metre length then we shortened the distance between the poles to make sure such detail was not missed.

The answer easily gets into level 2. It gives some detail and includes a reference to sampling but the answer is incomplete. It would be difficult for someone reading this account to carry out the technique from the information given. It is not clear exactly what is measured by the metre rule. A simple sketch might have helped here. The answer does identify a potential problem and how this can be overcome by shortening the distance between the vertical poles, but there is still the problem of exactly how the rise or fall is measured. The mark awarded would probably be just below full marks, in this case 5/6, credit being given for some reference to sampling, some detail of equipment used and the recognition of the need to adapt the technique at times.

Part (c)

Outline one technique that you used to present one aspect of your primary research and explain why it was appropriate. [6 marks]

Average response

I drew two long profiles of the wave-cut platform on three acetate sheets stuck together. Three sheets were needed because the profiles were so long. The acetate sheets meant I could make direct comparisons between the Wirral and Welsh sides of the island by laying one above the other. I chose a scale of one centimetre to one metre so I could show the slope of the profile and the changes in shape. The same scale was used on both axes so the shape would be realistic. The distance from the base of the cliff was shown on the horizontal axis and the rise and fall of the platform surface on the vertical axis.

The question is level marked.
The second part of the question is addressed at the start with an appropriate comment concerning the use of an overlay. Some details are given about the setting out and choice of scale, but these are not well developed or explained. Nevertheless the answer would just get into level 2 with a mark of 5.

Strong answer

A long profile was drawn for the wave-cut platform profile survey. This was drawn by hand on three pieces of graph paper stuck together. This was necessary because I did not want to use any vertical exaggeration on the vertical scale. I wanted the shape to be as realistic as possible. This meant that the vertical and horizontal scales needed to be the same. I chose a scale of one centimetre to one metre so I could fit the profile onto three pieces of graph paper and still show the slope of the profile and the changes in shape. I was able to draw profiles for the Welsh and Wirral sides one above the other which meant I could make direct comparisons and was very realistic since there was no exaggeration on the vertical scale. The long spread of the profile also meant that there was plenty of space to label clearly the detail I wanted to emphasize along the profile, such as the position and shape of the sudden scarps on the Welsh side and the places where there were blocks of rock or pebbles on the surface.

This is an excellent answer which shows a real understanding of the effect of choosing different scales on the appearance of a profile. It was quite a complicated process to explain but the student did it very well, probably because she had decided on what to do herself and then actually carried it out. There was detail given about the practicalities and the technicalities of the process. Note also that it is often worth presenting material 'by hand' rather than relying on the computer to produce something colourful but possibly inappropriate or not as useful. This answer easily gained the full 6 marks available.

Part (d)

To what extent did your findings support the geographical ideas which underpinned your investigation? **[5 marks]**

Average response

My aim was to show that the geology of the area had an influence on the coastal landforms of Hilbre Island. I was successful in showing this. Because the sandstone rocks were strong they formed cliffs and did not keep collapsing as some other local cliffs do when they are made out of boulder clay. There were also caves and stacks in places. One of the caves was produced when waves undercut the cliff because there was a very soft layer of marl at the foot of the cliff which was attacked by the waves at high tide. My wave-cut platform profiles showed that the Welsh side of the island was more rugged than the Wirral side, again due to the geology.

> The question is level marked.
> The candidate makes two points, both relating to the resistance to erosion of the rocks. There is some limited development but not enough to raise the answer to level 2. The reference to differences in the form of the wave-cut platform is not clearly linked to the geological structure, merely stated. The answer would get to the top of level 1 with 3 marks.
>
> Overall this candidate scored just 14/25.

Strong answer

I discovered that the geology of the area did have an effect on the cliffs and the wave-cut platform. The cliffs were vertical, or even overhanging in places due to the strong nature of the sandstone rock. This meant that when waves undercut the base of the cliff the rock above did not collapse straight away because it was strong enough to hold itself up. An unexpected point was that the cliffs on the Wirral side of the island were vertical or overhanging which is not what the textbooks suggest when the rocks are dipping towards the sea (the slope of the cliff should have been more gentle according to the textbook). These ideas would need further investigation.

My profiling of the wave-cut platform on either side of the island showed the influence of geological structure well. On the Welsh side the rocks were dipping in towards the cliff and the platform was irregular with some sections having a slope back towards the cliff. Again, this is not what the textbooks show. They usually show a wave-cut platform gradually sloping away from the foot of the cliff. On the Wirral side of the island, where the rocks were dipping towards the sea, the platform was smoother and had a slope which was always running down towards the sea.

> To get into level 2 the findings would need to be clearly linked to the aims of the enquiry and how these are linked to theory. The impact of particular characteristics of the local environment would need to be drawn out. These elements are clearly present in this answer. There is also a reference to further areas of research. It is not necessary to challenge accepted theory but when this is done, as here, it allows real understanding to be demonstrated and the complexity of the real world to be acknowledged. The full 5 marks would be awarded.
>
> Overall the candidate performed extremely well on this question, scoring 23/25.

Unit 2, Geog 2, Section 2: An alternative fieldwork investigation: human

You have experienced geography fieldwork as part of the course. Use this experience to answer the following questions.

Part (a) (i)
Describe the location of your fieldwork and outline why this was a suitable site for your investigation.
[**4 marks**]

Strong answer

We chose Morecambe for the location of our fieldwork on social deprivation. It is a seaside town, five miles from Lancaster town centre. Like many other seaside towns Morecambe has suffered decline due to lack of investment and cheap holidays to warm climates overseas. It has therefore become run down with many businesses having to close which has resulted in high unemployment. It was a good area to study the social deprivation because it has lots of decline and lots of areas with high unemployment. Morecambe has lots of derelict buildings and lots of areas with social deprivation. Morecambe is also a small town ☞

and therefore we could get a large spread of data given our time restraints. Transport is also good which means we could easily get around. It was also good to see the impact of regeneration schemes in Morecambe.

1 mark is awarded for each valid point. Location is clearly described (2 marks) and some detailed background is given describing the nature of the area and why it is appropriate for the study planned. This easily gains the other 2 marks available. If only 1 mark had been awarded for location the answer would still have gained full marks.

Part (a) (ii)
State one hypothesis or research question or issue for evaluation that you have investigated in 2(a) (i). Describe one method of primary data collection used in this investigation. [**5 marks**]

Strong answer

Does the West End of Morecambe have a higher level of social deprivation than Tarrisholme? We used environmental surveys to measure the social deprivation in Morecambe. We visited two super output areas in Morecambe, one in the West End and one in Tarrisholme. In each of these areas we used systematic sampling to mark 10 points in each area along a line of transect. We visited each of these points and conducted an environmental survey in each. The environmental survey consisted of a series of questions (such as state of repair). It featured a bi-polar scoring system where qualitative data could be ☞

converted into quantitative data and therefore we could give each point in each super output area a score.

The question is level marked. The hypothesis was clearly and simply stated and is linked to the specification. As soon as an appropriate method of data collection was identified and some detail given the answer would move into level 2. This answer goes on to give further detail about the application of the technique and makes an appropriate reference to sampling. The methodology was comprehensive and could be replicated. The full 5 marks would easily be gained.

Part (a) (iii)
Discuss the limitations of your chosen method in
2(a) (ii). [**6 marks**]

Average answer
The environmental survey featured lots of questions which were subjective in many parts. The sampling method that we chose could have inadvertently picked up a systematic bias in our results and therefore the houses that we surveyed might not have been representative of Morecambe. As well as this, large amounts of data can be missed which means that we didn't get a good spread of data. The environmental survey relied on qualitative data to measure the social deprivation which is hard to score and turn into quantitative data. As well as this, it was very time consuming to survey every house along every point. Lots of points marked by our systematic 👉

sampling method were not in residential areas and therefore our data might have been unrepresentative.

The question is level marked. To get into level 2 there needs to be some detail and a clear understanding of the limitations in the context of the investigation. It was judged that this answer lacked some detail and clarity of expression. It was rather simplistic and subjective in its discussion of limitations. The limitations could easily have been anticipated at the planning stage. It was given a mark just below the top of level 1, 3 marks out of the possible total of 6.

Part (b)
Outline and justify the use of one or more techniques
used to analyze your results. [**5 marks**]

Strong answer
I used a choropleth map to analyze my results. This involves marking the various scores for the social deprivation in different areas in Morecambe onto a map with different shading representing different values. This was a good method because it gave me a quick visual impression of the patterns of inequality in Morecambe and it showed me where the most and least deprived areas in Morecambe were in relation to the town centre. I also used a scattergraph to analyze my results. This uses two scales to plot environmental quality scores with distance from the town centre. I then drew 👉

a line of best fit. This showed me the general trend in the relationship and was quick and easy to produce and it allowed me to use all of my data. It also enabled me to identify the anomalous results in my data so that I could try to explain them.

The question is level marked. To get into level 2 there needs to be an appropriate choice of technique and some detail given. The choropleth detail would have just given access to level 2. The scattergraph section is stronger and would allow full marks to be awarded.

Part (c)
Drawing upon your findings, explain how your enquiry improved your understanding of the topic. [**5 marks**]

Strong answer

The investigation meant that I found out more about social deprivation and inequalities in an area that has suffered decline and unemployment. My results showed me that large areas in Morecambe have been affected by the decline of tourism in the area. Large areas were run down and houses were in poor condition as people don't have the money to maintain them. My findings also showed me that some areas were not affected, e.g. Tarrisholme which did not suffer from social inequality. This area had large well-kept houses and the residents were wealthy with large, neat gardens, showing that social deprivation does not affect all areas. The areas most affected were closer to the city centre of Morecambe and these have suffered most decline because the residents have become unemployed due to the closure of businesses. Areas that weren't affected were in the suburbs where the residents were not employed in the tourist industry and therefore were not affected and could maintain their houses. I learnt that social deprivation does not affect all areas but it does mean that people's quality of life is affected when they are unemployed and live in poor environments. People therefore suffer worse health in deprived areas.

The question is level marked. To get into level 2 there needs to be a clear link between the aim of the investigation, the findings and theory. This answer does just that and also shows an understanding of the special nature of the area studied. The final statement seems to be just tagged on at the end of the answer and is not supported. However, the full 5 marks were awarded.

Overall this was a very strong answer clearly linked to fieldwork carried out by the student.

Score for the question 22/25.

Unit 4A, Geog 4A, Section A: Geography fieldwork investigation

The answers relate to fieldwork carried out by a student in the context of the Hilbre Islands case study.

Part (a)

State the aim(s) of your investigation and explain the reasons why you selected this aim. [**10 marks**]

Strong answer

My aim was to investigate the effects of the approach of waves on the processes taking place and the landforms produced by marine erosion and deposition on and around Little Hilbre. This led on from my AS fieldwork carried out last year. I had discovered that there were lots of contrasts between the Welsh and Wirral sides of the island. The cliffs were lower on the Wirral side and the wave-cut platform was much smoother. On the Welsh side the cliffs were higher and the rock platform rugged. These contrasts were due to the nature of the sandstone rocks and the fact that they were dipping across the island. However, whenever I saw aerial photographs of the island or went onto the webcam camera site on the internet I could see that the waves always seemed to be approaching the islands from the northwest or north. I saw this for myself when I visited the island with my family in summer and stayed there over the high tide. We had done some work in our AS course on wave fetch and angle and how this was linked to longshore drift on sandy and pebbly beaches such as Spurn Head. I had not seen any descriptions of what happens on cliff coastlines and rock platforms when waves come to shore at an angle but I thought there must be some effect. ☞

I had seen that there was very little debris on the rock platform on the Wirral side of the island and thought that it might have been carried southeastwards over the rock platform which runs between Little Hilbre and Little Eye by a kind of longshore drift. So I decided to do two things.

- To survey the extent of loose clastic material on the wave-cut platform on either side of the island to see if this could be linked to longshore drifting.

- To investigate if any material moved southeastwards from Little Hilbre became smaller and more rounded because of abrasion and attrition.

This question is level marked, with 3 levels. The answer made clear links with previous work indicating ideas which could be further developed. These ideas appear to be well thought out and show some originality. This would ensure a mark at least within level 2. The links with previous work are quite strong and there are both theoretical and practical aspects to the suggestions made which should ensure a mark at the top of level 2. For level 3 more detail on the precise nature of the two aims or hypotheses might have been expected so a mark of 8/10 (top of level 2) would be awarded.

Part (b) *In the context of your chosen location, describe and justify the risk assessment you carried out.*
[**8 marks**]

Strong answer

When we prepared to visit the islands as a group our teacher had to complete a full risk assessment because this has to be done now for all trips. The school has a special form on which we had to identify all the possible hazards, assess the risks, giving a weighting to the degree of risk, and then suggest ways of overcoming or reducing the risks. We discussed the hazards and the level of risk and then did the following things. Tide tables were consulted because the islands can only be reached safely at low tide. We worked out the best time and how long we could stay out before returning. The tide comes in rapidly and we did not want to get trapped or have to walk through deep water. Because we were going out at low tide and the low water mark is some way from where we were going to be working the open sea was not a hazard and so there was no risk of drowning. We all planned to wear waterproof footwear with good gripping soles since the rock surfaces would probably be slippy after being uncovered by the tide. The initial visit was in late spring so there was little risk of exposure due to strong winds or low temperatures. However, we did plan to take warm waterproof clothing because there are few places to shelter and we checked the weather forecast the day before in case storms were forecast. Because we were working near cliffs we all had hard hats to wear in case of rock fall. We were also instructed to keep away from the base of ☞

the cliff and not to climb up. We were to work in threes and so if someone slipped or was hurt there was help at hand. Each group would have a mobile phone and we checked beforehand that there was reception on the island. It would be easy to just ignore these precautions but we were told of a previous geography field trip to the islands which had to be abandoned because one student did not take any warm clothing and became cold and could not stop shivering when a cold wind got up. Also the week before our second trip a woman slipped on the rocks, broke her leg and had to be taken off the island to hospital in the ranger's Land Rover.

The question is level marked. The candidate understands the difference between hazards and risks, and is aware of the steps in the risk assessment process. This would ensure level 2 is gained. The answer contains points relating to prior planning and specific actions on the visit. Specific detail relating to the fieldwork itself is given. Some basic ideas are missing, which may have been in the risk assessment, but were not referred to in this answer. For example, the need to inform the ranger of the visit, the need to take drinking water to avoid dehydration, and first-aid kit with sun cream to protect exposed skin. Nevertheless, it is still a very sound answer which does give justification for risk assessment. Full 8 marks.

Part (c) *Describe one data analysis technique you used and explain why this was suitable for your investigation.* [**10 marks**]

Average answer

I decided to use the Mann-Whitney U test to investigate the change in size of boulders and pebbles on the wave-cut platform which runs to the southeast of Little Hilbre towards Little Eye. I was expecting the material on the platform to become smaller as it is moved away from its source in the cliffs of Little Hilbre. I made the assumption that Little Hilbre was the main source of this material and that longshore drift would move material southeastwards, and that abrasion and attrition would reduce the size of the pebbles. At AS level I had used the Spearman test on pebbles on the rock platform close to the island and had got confusing results so I thought that with the Mann Whitney test there might be less confusion since I would only be taking measurements at either end of three lines drawn from Little Hilbre towards Little Eye. Collecting the data would also be quicker during the low tide period. Because I had actual values for pebble size at each data collection point the Mann Whitney test was better than chi-squared which uses grouped data categories.

For each of the three sites at both ends of the transect lines, the size of the longest axis of each of the 15 pebbles I measured was ranked in a table and the U value calculated from the formula. The results then had to be checked for significance using the critical values table for this test. Each of the three smaller values for U was checked in relation to the critical value at the 0.05 level of significance. In each case the smaller U value was lower than the critical value shown in the table. This meant that in 95 times out of 100 the difference between the medians for the data could not have occurred by chance. So I could reject my null hypothesis that there would be no significant 👉

difference in median sizes of the pebbles at either end of the transect lines. Therefore the pebbles had got smaller as I suspected and I could confidently put this down to attrition during the longshore drifting process.

The question is level marked. The link with the student's aim is made clear at the outset and reasons for choosing this test are given. Given the nature of the data the preference for Mann-Whitney over chi-squared is valid. The time consideration is also valid in the context of this location. However, the reason for not using Spearman is suspect. What is meant by 'confusing'? Does this just mean it did not give the results expected? Spearman may well have been a very good if not better alternative if time could have been given to data collection. Before applying the Mann-Whitney test it might have been useful to produce a simple dispersion graph to compare the data.

This is not an easy test to describe in words and the formula used is not an easy one to remember or explain without using actual measurements, which would be unavailable at the time of the examination. The second paragraph accurately describes how the test is run, followed by use of the critical values table and an explanation of the level of significance of the results. The assumptions then made are also possibly suspect in that the test does not prove anything and this should be made clear in any analysis. Based on the second paragraph, and with some credit being given to reasons for choice of technique, a high level 2 mark should be awarded. Because of the way Spearman was dismissed, and the uncritical acceptance of proof, a level 3 mark would not be appropriate. 7 marks.

Part (d) *Evaluate the success of your investigation in relation to your aims.* [**12 marks**]

Average answer

My investigation had two main parts to it. Both parts are concerned with the process of longshore drifting due to waves approaching from the north or northwest. First I wanted to survey the extent of loose clastic material on the wave-cut platform on either side of the island to see if this could be linked to longshore drifting. Then leading on from this to investigate if any material moved southeastwards from Little Hilbre became smaller and more rounded because of abrasion and attrition during the drifting process.

When I surveyed the loose material on the rock platform I found that there was very little of it on the Wirral side where the rocks were dipping away from the cliffs but on the Welsh side there was a lot of material at the foot of the cliffs and trapped below the small scarps across the platform. From this I deduced that the platform had been swept almost clear on the Wirral side by the longshore drifting due to waves approaching mostly from the open sea to the north. The dip of the rocks had added to this clearing of loose material because pebbles could be dragged down the smooth platform by the backwash. On the Welsh side the rock surface was sloping in towards the cliffs in places and this would mean a weaker backwash and that some of the pebbles would be trapped for a while before being moved southeastwards along the platform by the swash. My investigation of the pebbles which had

been moved southeastwards showed that they had become more rounded and smaller to the southeast. This was proved by my Mann-Whitney test of median size and my Cailleux index test for roundness. These also showed that longshore drift was taking place. Overall my results further developed my understanding by showing that the processes operating were the ones I had studied in my AS course but that in this location there were special considerations to take into account which made the landforms and some of the processes complex and special. For example, geology can influence the detailed landforms found in a coastal area, as was seen in the shape and slope of the wave-cut platform and how much material was on its surface. Also the fetch of prevalent waves can have an important effect on processes and landforms. The two acting together here further complicate the situation with the strong swash pushing material into the gulleys underneath the scarps and at the foot of the cliffs before moving the material along the rock platform in a southeasterly direction. For these reasons I feel my investigation was a success as I was able to prove what I set out to prove.

The question is level marked. The candidate makes a good attempt to describe and explain how the investigation came to its conclusions on the evidence gathered and how these conclusions link with the initial aims. The special characteristics of the area are emphasized and the link with the suggested processes clearly made. This should ensure a firm level 2 mark. However, there are assumptions made which ought to have been qualified. Longshore drifting has not been observed, measured or 'proven'. The evidence is circumstantial although perhaps quite persuasive if its limitations had been admitted. An additional weakness is that tests such as Mann-Whitney only give some statistical objectivity to a hypothesis – they do not prove that something has taken place. For these reasons the answer could not be moved up to level 3 and a mark of 8/12 in the middle of level 2 would be appropriate, because of limitations in terms of evaluation.

Overall this is a strong answer which gains 31/40.

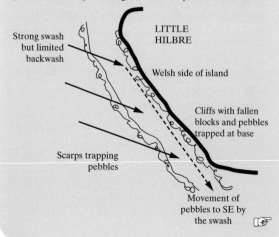

Strong swash but limited backwash

LITTLE HILBRE

Welsh side of island

Cliffs with fallen blocks and pebbles trapped at base

Scarps trapping pebbles

Movement of pebbles to SE by the swash ☞

Cartographical skills

Choropleth map	Shows relative density of a characteristic in an area. The density is represented by graded shading or line drawings
Density	The concentration of data or people in a particular location
Desire lines	Show the movement of people or traffic. The line is drawn from the point of origin to the destination. The width of the line is proportional to the volume of movement it represents
Distribution	The spread of data across an area
Dot maps	Show the spatial variation or density of a variable across an area. Each dot represents a certain value so the total values can be estimated by counting the dots
Flow lines	Show the quantity of movement of people or traffic along an actual route. The width of the line is proportional to the volume of movement it represents
Isoline	These are lines drawn on a map to represent points of equal value. There are many different types; for example, contour lines and isobars showing pressure
Lines of latitude	Imaginary lines that run horizontally (west–east) around the globe, including the equator
Lines of longitude	Imaginary lines that run vertically (north–south) around the globe, including the Greenwich Meridian Line
Proportional symbols	Placed on maps to show relative values in a data set. They can take the form of circles or squares. The size of the symbol is proportional to the value it represents
Trip lines	Straight lines drawn on a map to show journeys that people make, e.g. commuter destinations or origin of shoppers

Using graphical skills

Anomaly (residual)	These are points that lie away from the line of best fit. They don't fit the general trend or pattern but are useful as they may indicate other factors that affect the data
Bar graph	A graph whereby the length of the bars is proportional to the value of data they represent. They can be simple, with one set of data; compound, where the bars are divided into relative amounts; or comparative, where two or more sets of data are represented. Bar graphs are used for data with discrete values
Correlation	The relationship between two sets of data. It can be positive (as one set of data increases in value so does the other) or negative (as one data set increases the other decreases)
Cross-section	A slice cut through the landscape perpendicular to the surface, e.g. river or valley cross-section

Dependent variable (DV)	Data that is affected by the change in the other variable, e.g. velocity should increase because of distance downstream
Dispersion diagram	Used to show the distribution of data. The x-axis is narrow, often representing location or time, and the y-axis is longer, representing all the values in the data set
Independent variable (IV)	Data that is expected to cause the change in the other variable, e.g. distance downstream will affect velocity
Kite diagram	Shows changes in relative proportions of vegetation with distance from a point. One axis shows distance and the other axis shows plant species. The quantity of plant species is drawn either side of the axis so that it appears as a symmetrical pattern. Hence the term 'kite'
Line graph	Shows data points that are joined together by a line. It typically represents continuous data. Line graphs can be simple, compound, comparative or divergent
Logarithmic scale	A scale that is divided into cycles or intervals that increase 10-fold each time, e.g. on a Hjulstrom curve or on the Richter scale
Pie chart	A pie chart is a circle divided into sectors, each segment representing a share of the total value of the data set. In a pie chart, the angle of each sector is proportional to the quantity it represents
Proportional divided circles	Similar to pie charts with segments representing the share of the data set, but in this case the size of the circle is proportional to the total value of the data set
Radial diagram	Used for data that has a directional or temporal component. The frequency or percentage is recorded on the radius of the diagram. The circumference varies according to the data; for example, compass points for wind direction
Scattergraph	Used to show the relationship between two variables. A line of best fit is often added to represent the trend
x-axis	The horizontal axis
y-axis	The vertical axis

Statistical skills

Central tendency	A measure of the middle value of a set of data. There are various ways of measuring central tendency: mean, mode and median
Chi-squared test (X^2)	A comparative statistical technique that compares observed data against theoretical data (expected data)
Critical value	Used in conjunction with many statistical techniques to test the significance level. Each technique will have a table of critical values against which the result of the statistical test is compared
Dispersion	The spread of data in a set of values
Expected data	A theoretical distribution of data

Interquartile range (IQR)	The difference between the upper quartile and the lower quartile. The extreme values are therefore removed, and only the middle 50% of data is considered
Mann-Whitney U test (U)	A statistical technique that tests to see if there is a difference between the medians of two sets of data
Mean	Sometimes called the average, is calculated by adding up all the values in a data set and dividing the total sum by the number of values in the data set
Median	The middle value in a data set that has been rank ordered
Mode	The most frequently occurring value in a data set
Non-parametric test	A statistical test that assumes the data is not normally distributed
Normal distribution	A theoretical frequency distribution, usually in a bell-shaped curve that is symmetrical about the mean
Null hypothesis	Many statistical techniques are based on testing the assumption of a null hypothesis. A null hypothesis is usually: 'There is no relationship between variable 1 and variable 2'
Observed data	Actual data collected from a primary source or a secondary source
Range	The difference between the highest and lowest value in a data set
Significance level	This is the level at which you can be confident that your results did not occur by chance. Usually the 95% and 99% levels are used
Spearman's rank cor-relation test	A statistical technique that tests the strength of a relationship between two variables. The result (R_s) always lies between –1 and +1
Standard deviation	A measure of how spread the data is about the mean

Fieldwork at AS and at A2

Clasts (clastic fragments)	The general geological term for loose rock fragments broken from solid rock. Depending upon size, shape and location they may be more commonly referred to as blocks, pebbles, shingle or sand
Clinometer	An instrument for measuring angles in the field. It can be a very simple plastic pivoting pointer or a sophisticated digital sighting instrument
Current bedding	Sloping layers of rock running at an angle to the general bedding in some sedimentary rocks such as sandstones. The original deposition would have been in desert dunes or river deltas
Dip (of rocks)	The angle (measured in degrees) which a layer of rock (usually a sedimentary rock) makes with the horizontal
Fault	A physical displacement of adjacent rock strata due to tectonic movement. The displacement may be hundreds or even thousands of metres. Despite this, most faults are not clearly visible at the earth's surface

Joints	Visible cracks (frequently more or less vertical) running through rock bedding. Joints, unlike faults, show no vertical displacement of bedding but often provide access for water penetration and are zones of weakness which can be exploited by weathering and erosion processes
Neighbourhood statistics	Groups of census and other statistics available through the ONS. They are particularly useful when investigating aspects of urban geography
Office for National Statistics (ONS) (www.ons.gov.uk)	The ONS produces an array of statistics providing information on the state of society and the economy across the United Kingdom. For geographers the key statistics for which the ONS is responsible are: the census; labour market statistics on jobs and employment; population and demography data including emigration and migration; economic and social surveys; regional and neighbourhood statistics
Ranging pole	A simple, but useful, piece of surveying equipment, usually 2 metres in length and painted alternately red and white in half-metre sections. It can be used in levelling surveys in conjunction with simple or advanced levelling instruments
Risk assessment	A structured process of assessing hazards and risks which should be carried out when fieldwork is planned. The process involves identifying the hazards, deciding who might be affected, assessing the likelihood and severity of the risk, taking measures to eliminate or reduce the risk
Super Output Areas (SOAs)	SOAs are small areas specifically introduced to improve the reporting and comparison of local statistics. Within England and Wales there is a lower layer (minimum population 1000) and a middle layer (minimum population 5000). Unlike electoral wards, these SOA layers are of consistent size across the country and are not subjected to regular boundary change
Ward	Electoral wards are the base unit of UK administrative geography, being the areas from which local authority councillors are elected. They are commonly referred to simply as wards. Some ONS statistics are presented at this level, for example the total deprivation diagram on page 56

Geographical issue evaluation

Cost-benefit analysis	A formal way of analyzing a planned course of action or a planned development. The financial, social and environmental costs or repercussions of the proposed action are weighed against the benefits which are likely to be experienced. Precise financial costs and gains can sometimes be calculated but social and environmental considerations are more difficult to quantify. Opportunity costs are often associated with such analyses

Economic sectors	The term 'sector' can be used in different ways in geography. Here, it is concerned with the ownership of organizations. The private sector includes commercial businesses with the primary objective of making profits for the owners or shareholders. The **public sector** includes non-profit making bodies set up by governments. Their primary objective is to provide services and carry out government policy. **Voluntary sector** organizations are also set up to provide services but are independent of government, although they may be supported by, and work with, government in providing services
NIMBY ('not in my back yard')	A term applied in situations where a proposed development may be needed, and generally recognized as being for the common good, but people living close by, who may be affected, object on personal grounds
Non-governmental organisations (NGOs)	Legally constituted organizations which operate independently from any government and are regarded as being politically impartial. Many are, essentially, charities involved in social or environmental work. They can have a political aspect to them but are not directly set up or ultimately responsible to any government or political party. They can, however, be at least partly funded by governments and can help governments to carry out their programmes. Many operate on a national level, such as the National Trust. Others are global in their activities, such as the Red Cross.
Pressure (interest) groups	Groups of like-minded individuals or organizations hoping, through the power of numbers and argument, to influence policy and political decision making. They can form and operate on a small local scale, nationally or globally. A number of NGOs effectively act as pressure groups; for example, Greenpeace and Tourism Concern
Stakeholders	A general term given to individuals, groups or organizations having a vested interest in a specific plan, development or issue